Optimization Algorithms for Networks and Graphs

INDUSTRIAL ENGINEERING

A Series of Reference Books and Textbooks

Editor

WILBUR MEIER, JR.
Head, School of Industrial Engineering
Purdue University
West Lafayette, Indiana

Volume 1: Optimization Algorithms for Networks and Graphs, *Edward Minieka*

Additional Volumes in Preparation

Optimization Algorithms for Networks and Graphs

Edward Minieka

Department of Quantitative Methods
University of Illinois at Chicago Circle
Chicago, Illinois

MARCEL DEKKER, INC.   New York and Basel

Library of Congress Cataloging in Publication Data

Minieka, Edward.
 Optimization algorithms for networks and graphs.

 (Industrial engineering; v.1)
 Includes bibliographical references and index.
 1. Graph theory. 2. Network analysis (Planning).
 3. Algorithms.   I. Title. II. Series.
 QA166.M56   519.7   77-29166
 ISBN 0-8247-6642-3

COPYRIGHT © 1978 by MARCEL DEKKER, INC.
ALL RIGHTS RESERVED

Neither this book nor any part may be reproduced or transmitted in any form or by any means, electronic or mechanical, including photocopying, microfilming, and recording, or by any information storage and retrieval system, without permission in writing from the publisher.

MARCEL DEKKER, INC.
270 Madison Avenue, New York, New York 10016

Current printing (last digit):
10 9 8 7 6 5 4 3 2

PRINTED IN THE UNITED STATES OF AMERICA

To Eva and Stanley,
who would have been pleased

PREFACE

This is not another graph theory text; it is a text about algorithms--optimization algorithms for problems that can be formulated in a network or graph setting.

As a thesis student studying these algorithms, I became aware of their variety, elegance, and interconnections and also of the acute need to collect and integrate them between two covers from their obscure hiding places in scattered journals and monographs. I hope that I have been able in the confines of this text to make a stab at a comprehensive, cohesive, and clear treatment of this body of knowledge.

This text is self-contained at the level of an advanced undergraduate or beginning graduate student in any discipline. An operations research or mathematics background is, of course, helpful but hardly essential.

The text aims at an intuitive approach to the inner workings, interdependencies, and applications of the algorithms. Their place in the hierarchy of advanced mathematics and the details of their computer coding are not stressed.

Chapter 1 contains background information and definitions. Aside from Chapter 1, I have tried to make all chapters as independent of one another as possible, except where one algorithm uses another as a subroutine. Even in this situation, the reader can continue if he is willing to accept the subroutine algorithm on faith. This text can be treated comprehensively in a one-semester course and less thoroughly, but without significant omissions, in a one-quarter course.

It takes a lot of ink to go from source to sink. I wish to thank Randy Brown at Kent State University, Ellis Johnson at IBM, George Nemhauser at Cornell University, and Douglas Shier at the National Bureau of Standards for their careful readings and suggestions. Also, emphatic thanks to my colleague Leonard Kent for his confidence in this project and for years of encouragement. Lastly, thanks to my students who graciously endured three years of classroom testing of the various manuscript stages. Most of all, this book is for you and for your successors, everywhere.

Edward Minieka

CONTENTS

PREFACE v

1. INTRODUCTION TO GRAPHS AND NETWORKS 1

   1.1 Introduction 1
   1.2 Some Concepts and Definitions 4
   1.3 Linear Programming 8
   Exercises 15
   References 16

2. TREE ALGORITHMS 19

   2.1 Spanning Tree Algorithms 19
   2.2 Maximum Branching Algorithms 26
   Exercises 36
   References 39

3. PATH ALGORITHMS 41

   3.1 Shortest Path Algorithm 41
   3.2 All Shortest Path Algorithms 51
   3.3 The K-th Shortest Path Algorithm 64
   3.4 Other Shortest Paths 78
   Exercises 81
   References 84

4. FLOW ALGORITHMS 87

   4.1 Introduction 87
   4.2 Maximum Flow Algorithm 95
   4.3 Minimum Cost Flow Algorithm 105
   4.4 Out-of-Kilter Algorithm 116
   4.5 Dynamic Flow Algorithms 128
   4.6 Flows with Gains 151
   Exercises 174
   References 179

5. MATCHING AND COVERING ALGORITHMS 181

   5.1 Introduction 181
   5.2 Maximum Cardinality Matching Algorithm 185
   5.3 Maximum Weight Matching Algorithm 200

	5.4 Minimum Weight Covering Algorithm	214
	Exercises	231
	References	233
6.	**POSTMAN PROBLEM**	**235**
	6.1 Introduction	235
	6.2 Postman Problem for Undirected Graphs	238
	6.3 Postman Problem for Directed Graphs	245
	6.4 Postman Problem for Mixed Graphs	249
	Exercises	258
	References	260
7.	**TRAVELING SALESMAN PROBLEM**	**261**
	7.1 Salesman Problems	261
	7.2 Existence of a Hamiltonian Circuit	265
	7.3 Lower Bounds	272
	7.4 Solution Techniques	277
	Exercises	283
	References	286
8.	**LOCATION PROBLEMS**	**289**
	8.1 Introduction	289
	8.2 Center Problems	298
	8.3 Median Problems	306
	8.4 Extensions	315
	Exercises	316
	References	318
9.	**PROJECT NETWORKS**	**319**
	9.1 Critical Path Method (CPM)	319
	9.2 Minimum Cost Activity Times	333
	9.3 Generalized Project Networks	342
	Exercises	349
	References	352
	INDEX	**353**

Optimization Algorithms
for Networks and Graphs

Chapter 1

INTRODUCTION TO GRAPHS AND NETWORKS

1.1 INTRODUCTION

Graph theory is a branch of mathematics that has wide practical application. Numerous problems arising in such diverse fields as psychology, chemistry, electrical engineering, transportation planning, management, marketing, and education can be posed as problems in graph theory. Because of this, graph theory is not only an area of interest in its own right but also a unifying basis from which the results from other fields can be collected, shared, extended, and disseminated.

Unlike other scientific fields, graph theory has a definite birthday. The first paper on graphs was written by the Swiss mathematician Leonhard Euler (1707-1783) and was published in 1736 by the Academy of Science in St. Petersburg. Euler's study of graphs was motivated by the so-called Konisberg bridge problem. The city of Konigsberg (now called Kaliningrad) in East Prussia was built at the junction of two rivers and the two islands formed by them (see Fig. 1.1). In all, there were seven bridges connecting the islands to each other and to the rest of the city. Could a Konigsberger start from his home and cross each bridge exactly once and return home? The answer is no, and we will see why later (Chapter (Chapter 6) when we study a generalized version of this problem called the postman problem.

The growth of graph theory continued in the late nineteenth and early twentiety centuries with advances motivated by molecular theory and electrical theory. By the 1950s, the field had taken two essentially different directions: the *algebraic* aspects of graph theory and the *optimization* aspects of graph theory. The latter was greatly advanced by the advent of the computer and the discovery of linear programming techniques. This text is concerned almost exclusively with the optimization aspects of graph theory.

What is a graph? A graph consists of two parts, points and arrows joining these points. The points can be depicted as points in a plane or, if you prefer, points without any specific physical location. The arrows can be depicted as lines (either straight or curved) joining pairs of points. For example, the points of the graph shown in Fig. 1.2 are a, b, c, d, and the arrows of this graph are α, β, γ, δ, ϵ, ϕ. Notice that there are two arrows α and β that go from point a to point b, i.e., with the tail at a and the head at b. The same graph could be specified without using a

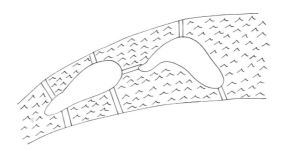

Figure 1.1
Konigsberg in 1736

picture simply by listing its points a, b, c, d and listing its arrows $\alpha = (a,b)$, $\beta = (a,b)$, $\gamma = (c,a)$, $\delta = (b,c)$, $\epsilon = (b,d)$, $\phi = (d,c)$ as ordered pairs of points, where the first point in an ordered pair denotes the point at the arrow's tail and the second point in an ordered pair denotes the point at the arrow's head. In keeping with standard terminology, we shall refer to the points of a graph as *vertices* and the arrows of a graph as *arcs*. With this as motivation, we can now state a formal definition for a graph:

1.1 Introduction

A *graph* is a set X whose numbers are called *vertices* and a set A of ordered pairs of vertices. The members of A are called *arcs*, and the graph is denoted by (X,A).

Throughout, we shall assume that both set X and set A contain only a finite number of members.

In general, vertices will be denoted by small roman letters, and arcs will be denoted by small Greek letters or as ordered pairs of vertices. For example, in Fig. 1.2, arc γ can be denoted by (c,a), where vertex c is at the tail of arc γ and vertex a is at

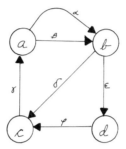

Figure 1.2

the head of arc γ. If there is more than one arc going from one vertex to another vertex, then each of these arcs can be denoted as a subscripted ordered pair of vertices. For example, in Fig. 1.2, arc α could be denoted by $(a,b)_1$ and arc β could be denoted by $(a,b)_2$. When no confusion will develop, we shall omit this subscript

EXAMPLE 1. Let X denote the set of all airports in Illinois. Let A denote the set of all pairs of airports (x,y) such that there is a nonstop commercial flight from airport x to airport y. Clearly, (X,A) is a graph.

EXAMPLE 2. Let X denote the set of all passengers aboard a certain transatlantic flight. Let A denote the set of all pairs (x,y) of passengers such that passenger x is older than passenger y and both speak a common language. Clearly, (X,A) is a graph with vertex set X and arc set A. Is it possible for this graph to have both an arc (x,y) and also an arc (y,x)?

Certain problems in graph theory, like the simple matching problem that we shall encounter later, require only a knowledge of the endpoints of each arc. In cases like this, the head and tail of each arc need not be specified, or in other words, the *direction* of each arc need not be specified. A graph whose arc directions are not specified is called an *undirected* graph. An undirected arc is called an *edge*. For example, if the arrow heads were removed from the graph in Fig. 1.2, the resulting graph would be an undirected graph, and the arcs would be called edges.

Throughout this text, we will use the notation (X,E) to denote an undirected graph with vertex set X and edge set E, and we will use the notation (X,A) to denote a graph with vertex set X and arc set A.

A *network* is merely a graph with one or more numbers associated with each arc. For example, if the air mileage were associated to each arc in Example 1, then this graph would be a network.

Each remaining chapter of this text is devoted to a single type of practical optimization problem on a graph or network. The procedures that are given for finding numerically optimum solutions to these problems are called *algorithms*, hence, the title of this text. As some of these optimization algorithms build upon others the order of presentation is restricted, and these considerations have dictated the sequencing of the chapters of this text.

1.2 SOME CONCEPTS AND DEFINITIONS

To decrease the dependence between chapters, some basic concepts and definitions that are needed throughout are presented here. The motivation for these definitions will be reserved to later chapters where applications are discussed in greater depth.

An arc that has the same vertex for both its head and its tail is called a *loop*. In Fig. 1.3, arc β is a loop.

A vertex and an arc are said to be *incident* to one another if the vertex is an endpoint (either the head or the tail) of the arc. In Fig. 1.3, arc α and vertex b are incident to one another.

1.2 Some Concepts and Definitions

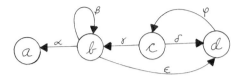

Figure 1.3

Two arcs are said to be *incident* to one another if they are both incident to the same vertex. In Fig. 1.3, arcs α and γ are incident to one another since they are both incident to vertex b.

Two vertices are said to be *adjacent* to one another if there is an arc joining them. In Fig. 1.3, vertices b and c are adjacent to one another since there is an arc γ that joins them.

Consider any sequence $x_1, x_2, \ldots, x_n, x_{n+1}$ of vertices. A *chain* is any sequence of arcs $\alpha_1, \alpha_2, \ldots, \alpha_n$ such that the endpoints of arc α_i are x_i and x_{i+1} for $i = 1, 2, \ldots, n$. Thus, either $\alpha_i = (x_i, x_{i+1})$ or $\alpha_i = (x_{i+1}, x_i)$. Vertex x_1 is called the *initial vertex* of the chain. Vertex x_{n+1} is called the *terminal vertex* of the chain. The chain is said to extend from its initial vertex to its terminal vertex. The *length* of a chain equals the number of arcs in the chain. In Fig. 1.3, the sequence $\alpha, \beta, \gamma, \delta, \phi$ of arcs forms a chain of length 5 from vertex a to vertex c.

A *path* is a chain for which $\alpha_i = (x_i, x_{i+1})$ for $i = 1, 2, \ldots, n$. The length, initial vertex, and terminal vertex of a path can be defined similarly. For example, in Fig. 1.3 the arcs β, ε, ϕ form a path of length 3 from vertex b to vertex c.

A *cycle* is a chain whose initial vertex and terminal vertex are identical. A *circuit* is a path whose initial vertex and terminal vertex are identical. The length of a cycle or a circuit is defined as the length of the corresponding chain. For example, in Fig. 1.3, arcs $\gamma, \varepsilon, \delta$ form a cycle of length 3, and arcs $\gamma, \varepsilon, \phi$ form a circuit of length 3.

A chain, path, cycle, or circuit is called *simple* if no vertex is incident to more than two of its arcs (i.e., if the chain, path, cycle, or circuit properly contains no cycles). In Fig. 1.3, chain α, γ is simple whereas chain α, β, γ is not; cycle $\gamma, \varepsilon, \delta$ is simple but cycle $\gamma, \beta, \varepsilon, \delta$ is not.

6   Introduction to Graphs and Networks

A graph is called *connected* if there is a chain joining every pair of distinct vertices in the graph. For example, the graphs of Figs. 1.2 and 1.3 are connected, but the graph in Fig. 1.4 is not connected because there is no chain joining vertices d and e. A graph may be regarded as consisting of a set of connected graphs. Each of these connected graph is called a *component* of the original graph. The graph in Fig. 1.4 has two components. (What are they?)

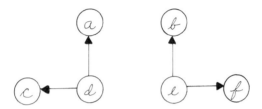

Figure 1.4

Let X' be any subset of X, the vertex set of graph G = (X,A). The graph whose vertex set is X' and whose arc set consists of all arcs in A with both endpoints in X' is called the *subgraph generated by X'*.

Let A' be any subset of A the arc set of graph G = (X,A). The graph whose arc set is A' and whose vertex set consists of all vertices incident to an arc in A' is called the *subgraph generated by A'*. For example, in Fig. 1.3 the subgraph generated by the vertices {a,b,c} is shown in Fig. 1.5. The subgraph generated by the arcs {γ,δ,φ} is shown in Fig. 1.6.

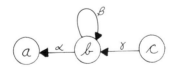

Fig. 1.5
Subgraph Generated by {a,b,c}

1.2 Some Concepts and Definitions

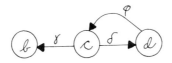

Fig. 1.6

Subgraph Generated by $\{\gamma, \delta, \phi\}$

A set of arcs is called a *tree* if it satisfies two conditions:
1. The arcs generate a connected subgraph
2. The arcs contain no cycles.

In Fig. 1.3, the following sets of arcs each form a tree:

$\{\alpha,\gamma,\varepsilon\}$, $\{\alpha,\gamma,\phi\}$, $\{\alpha,\varepsilon,\delta\}$, $\{\phi,\gamma\}$, $\{\alpha,\gamma\}$, $\{\varepsilon\}$, $\{\gamma\}$.

The arcs $\{\phi,\gamma,\varepsilon\}$ do not form a tree since they contain a cycle.

A *forest* is any set of arcs that contains no cycles. Thus, a forest consists of one or more trees. The arcs in Fig. 1.4 form a forest consisting of two trees.

A *spanning tree* of a graph is any tree formed from the arcs of the graph that includes every vertex in the graph. In Fig. 1.3, the arcs $\{\alpha,\varepsilon,\phi\}$ form a spanning tree since they include each vertex a,b,c,d. Clearly, no spanning tree can exist for a graph with more than one component, and every connected graph possesses a spanning tree.

A tree with one arc contains two vertices; a tree with two arcs contains three vertices; a tree with three arcs contains four vertices, and in general, a tree with n - 1 arcs must contain n vertices. Hence, each spanning tree of a (connected) graph with n vertices consists of n - 1 arcs.

A set of arcs whose removal from the graph increases the number of components in the graph is called a *cut*. A cut that contains as a proper subset no other cut is called a *simple cut*. In Fig. 1.3, arcs $\{\delta,\varepsilon,\gamma,\phi\}$ form a cut since their removal from the graph would yield a graph with 3 components. Also, arcs $\{\delta,\varepsilon,\phi\}$ form a cut since their removal from the graph would yield a graph with 3 components. Cut $\{\delta,\varepsilon,\gamma,\phi\}$ is not a simple cut since it contains another cut $\{\delta,\varepsilon,\phi\}$, which happens to be a simple cut.

Let G be any loopless graph with m vertices and n arcs. Let $\underline{G}$ be a matrix with m rows, one for each vertex, and n columns, one for each arc. Let $\underline{G}_{ij}$ denote the element in the i-th row and j-th column and be defined as follows:

$$\underline{G}_{ij} = \begin{cases} +1 & \text{if the vertex associated with row i is the head of the arc associated with column j} \\ -1 & \text{if the vertex associated with row i is the tail of the arc associated with column j} \\ 0 & \text{otherwise.} \end{cases}$$

Thus, each column of matrix $\underline{G}$ contains all zeros except for a +1 and a -1. Matrix $\underline{G}$ is called the *matrix of graph G*. The matrix associated with the graph in Fig. 1.2 is

	α	β	γ	δ	ε	φ
a	-1	-1	+1	0	0	0
b	+1	+1	0	-1	-1	0
c	0	0	-1	+1	0	+1
d	0	0	0	0	1	-1

Obviously, a matrix can be associated to a graph if, and only if, every column of the matrix contain all zeros except for a +1 and a -1.

Matrix representation provides a convenient way to describe a graph without listing vertices and arcs or drawing pictures. Computer programs for the optimization algorithms described in this text invariably use the matrix representation. However, in the interest of easing the presentation of the material which follows, all graphs in this text are presented in the more intuitive pictorial form.

1.3 LINEAR PROGRAMMING

Many of the problems considered in subsequent chapters can be reformulated as linear programming problems. This section gives a brief review of the linear programming results that will be needed in subsequent chapters. Although the results given in this section are sufficient for a formal understanding of the material in subsequent chapters, the presentation here is by no means intended to

1.3 Linear Programming

produce a profound or intuitive understanding of linear programming. For an intensive treatment of linear programming results, the reader is referred to any standard text on linear programming.

Let $x_1, x_2, \ldots, x_n$ be decision variables that can assume any non-negative real value. Let $c_1, c_2, \ldots, c_n, b_1, b_2, \ldots, b_m, a_{ij}$, be real numbers for $i = 1, 2, \ldots, m$ and $j = 1, 2, \ldots, n$.

A linear programming problem is any problem that can be fitted into the following format:

Maximize

$$\sum_{j=1}^{j=n} c_j x_j \tag{1}$$

such that

$$\sum_{j=1}^{j=n} a_{1j} x_j \leq b_1$$

$$\sum_{j=1}^{j=n} a_{2j} x_j \leq b_2 \tag{2}$$

$$\cdots \cdots$$

$$\sum_{j=1}^{j=n} a_{mj} x_j \leq b_m$$

$$x_1 \geq 0, \quad x_2 \geq 0, \quad \ldots, \quad x_n \geq 0 \tag{3}$$

Moreover, the $\leq$ sign in any relation (2) can be replaced by a $\geq$ sign or an = sign. Also, the $\geq$ sign in any relation (3) can be replaced by a $\leq$ sign, or the relation can be omitted entirely, in which case there is no restriction on the decision variable. Lastly, in expression (1), the "maximize" can be changed to "minimize." The following problem fits into the linear programming format:

Maximize

$$2x_1 + 7x_2 - 3x_3$$

such that

$$2x_1 - 2x_2 + 1x_3 \leq 6$$

$$4x_1 - 6x_2 - 3x_3 \geq 7.325$$

$-8.25x_1 + 1x_2 - 0.3x_3 = 8$

$x_1 \geq 0$, $x_2 \leq 0$, x_3 (unrestricted).

The expression in relation (1) that is to be maximized or minimized is called the *objective function*. The objective function is a linear combination of the decision variables x_1, x_2, ..., x_n. The m expressions of relation (2) that must be satisfied by the d decision variables are called the *constraints* of the linear programming problem. Note that the left side of each constraint is a linear combination of the decision variables. The expressions of (3) are called the *nonnegativity conditions* of the linear programming problem.

Given a linear programming problem with n decision variables and m constraints in the form of (1), (2), and (3), we can generate from it another linear programming problem called its *dual*. The dual linear programming problem has n constraints, one corresponding to each decision variable in the original problem; and the dual has m decision variables y_1, y_2, ..., y_m, one corresponding to each original constraint. The dual linear programming problem for the linear programming problem in (1), (2), and (3) is

Minimize
$$\sum_{i=1}^{i=m} b_i y_i \tag{1'}$$

such that

$$\sum_{i=1}^{i=m} a_{i1} y_i \geq c_1$$

$$\sum_{i=1}^{i=m} a_{i2} y_i \geq c_2$$

. . . . (2')

$$\sum_{i=1}^{i=m} a_{in} y_i \geq c_n$$

$y_1 \geq 0$, $y_2 \geq 0$, ..., $y_m \geq 0$ (3')

If the i-th original constraint in (2) was a $\geq$ inequality instead of a $\leq$ inequality, then the i-th nonnegativity constraint in the dual

1.3 Linear Programming

(3') is $y_i \leq 0$. If the i-th original constraint in (2) was an equality instead of an inequality, then the i-th nonnegativity constraint in the dual (3') is omitted and y_i is not restricted in sign.

If the j-th original nonnegativity constraint is $x_j \leq 0$, then the corresponding j-th dual constraint (2') is a $\leq$ inequality, that is, $\Sigma a_{ij} y_i \leq c_j$. If x_j is unrestricted in sign, then the j-th dual constraint is an equality constraint, that is, $\Sigma a_{ij} y_i = c_j$.

Lastly, if the original linear programming problem is a maximize problem, then its dual is a minimize problem. If the original linear programming problem is a minimize problem, then its dual is a maximize problem.

The original linear programming problem is usually denoted as the *primal*. Since the dual is also a linear programming problem, linear programming problems can be regarded as coming in primal-dual pairs. Moreover, the reader can easily show that the dual of a dual linear programming problem is the original primal.

The dual to the above linear programming problem is

Minimize

$$-6y_1 + 7.325y_2 + 8y_3$$

such that

$$2y_1 + 4y_2 - 8.25y_3 \geq 2$$

$$-2y_1 - 6y_2 + 1y_3 \leq 7$$

$$1y_1 - 3y_2 - 0.3y_3 = -3$$

$$y_1 \geq 0, \quad y_2 \leq 0, \quad y_3 \text{ (unrestricted)}.$$

How can we determine if a solution $x_1, x_2, \ldots, x_n$, that satisfies all the relations in (2) and (3), optimizes the objective function (1)? Also, how can we determine if a dual solution $y_1, y_2, \ldots, y_m$, that satisfies all relations (2') and (3'), optimizes the objective function (1')? In other words, how can we determine when a feasible linear programming solution is an optimal solution?

An answer to these questions is provided by a result from linear programming theory called the *complementary slackness conditions*.

Let $x_1, x_2, \ldots, x_n$ be a set of feasible values for the primal decision variables. Let $y_1, y_2, \ldots, y_m$ be a set of feasible values for the dual decision variables. Then, from equations (2) and (2'), it follows that the primal objective function equals

$$\sum_{i=1}^{i=n} c_i x_i \leq \sum_{i=1}^{i=n} x_i \sum_{j=1}^{j=m} a_{ji} y_j = \sum_{j=1}^{j=m} y_j \sum_{i=1}^{i=n} a_{ji} x_i \leq \sum_{j=1}^{j=m} y_j b_j \quad (4)$$

which is the dual objective function. Hence the primal objective function will always take a value less than or equal to the value of the dual objective function. If values could be found for the primal and dual decision variables such that equality held throughout equation (4), then those values would be optimal values for the primal and dual decision variables. Hence, we must seek conditions under which relation (4) holds with equality throughout.

The left inequality in relation (4) holds with equality if, and only if,

$$\left(\sum_{j=1}^{j=m} a_{ji} y_j - c_i\right) x_i = 0 \quad (i = 1, 2, \ldots, n). \quad (5)$$

The right inequality in relation (4) holds with equality if, and only if,

$$\left(b_j - \sum_{i=1}^{i=n} a_{ji} x_i\right) y_j = 0 \quad (j = 1, 2, \ldots, m) \quad (6)$$

Hence, if a set of feasible values for the primal and dual also satisfy equations (5) and (6), then these values for optimal values for their respective linear programming problems. Equations (5) and (6) provide a way to determine whether feasible solutions are optimal for a primal-dual pair and why they are called the *complementary slackness conditions*.

The difference $\sum_{j=1}^{j=m} a_{ji} y_j - c_i$ in equation (5) is called the *slack in the i-th dual constraint*. Likewise, the difference

1.3 Linear Programming

$b_j - \sum_{i=1}^{i=n} a_{ji}x_i$ in equation (6) is called the *slack in the j-th primal constraint*.

A linear programming problem in which some of the constraints are inequalities can be converted into an equivalent linear programming problem with all equality constraints in the following way.

If the i-th constraint is $\Sigma a_{ij}x_j \leq b_i$, add a new decision variable $s_i \geq 0$ to form a new constraint $\Sigma a_{ij}x_j + s_i = b_i$. Let the coefficient of s_i in the objective function be zero. This leaves the objective function unchanged and merely converts the i-th constraint into an equality constraint.

If the i-th constraint is $\Sigma a_{ij}x_j \geq b_i$, subtract a new decision variable $s_i \geq 0$ to form a new constraint $\Sigma a_{ij}x_j - s_i = b_i$. Let the coefficient of s_i in the objective function be zero. This leaves the objective function unchanged and merely converts the i-th constraint into an equality constraint.

For example, the linear programming problem in the preceding discussion could be converted into the following equivalent linear programming problem with all equality constraints:

Maximize

$$2x_1 + 7x_2 - 3x_3 + 0s_1 + 0s_2$$

such that

$$2x_1 - 2x_2 + 1x_3 + 1s_1 + 0s_2 = 6$$

$$4x_1 - 6x_2 - 3x_3 + 0s_1 - 1s_2 = 7.325$$

$$-8.25 + 1x_2 - .3x_3 + 0s_1 + 0s_2 = 8$$

$$x_1 \geq 0, \quad x_2 \leq 0, \quad s_1 \geq 0, \quad s_2 \geq 0, \quad x_3 \text{ (unrestricted)}$$

In general, a linear programming problem with all equality constraints will have more decision variables than constraints, that is, $n > m$. Thus, there will be $n - m$ more variables than constraints. Suppose that we arbitrarily select $n - m$ of the decision variables and decree that these variables equal zero. Then, these $n - m$ variables could be removed from the constraints. The resulting set

of constraints would consist of m linear equations in m unknowns.
This set of m simultaneous linear equations could be solved by any
standard method such as Cramer's rule or the Gauss-Jordan elimination
method. If a unique solution exists in which all m remaining variables
take values greater than or equal to zero, then this solution is called
a *basic solution*. Since there can be at most one basic solution for
each choice of n - m variables initially set equal to zero, there
can only be a finite number of distinct basic solutions.

One of the most important results of linear programming theory
is that

> If there is at least one optimal solution to a linear programming
> problem, there exists an optimal solution that is also a basic
> solution.

Hence to find an optimal solution, we need only examine a finite
number of basic solutions.

In practice, linear programming problems are solved by a method
called the *simplex algorithm*, which starts with one basic solution
and then judiciously generates another basic solution with a better
value for the objective function. This process is repeated until a
basic solution that is recognized as optimal is found. This process
is repeated only a finite number of times since no basic solution
reappears (since at each step a better solution is generated), and
there are only a finite number of distinct solutions.

Some of the graph problems presented in subsequent chapters
will be solved in a similar manner, namely, a basic solution will
be found and then another basic solution that is better will be
generated from the original basic solution. This process will be
repeated a finite number of times until an optimal solution is
discovered.

EXERCISES

1. Construct a graph whose vertex set is the set of courses you are required to pass for your degree. Place an arc from vertex x to vertex y if course x is a prerequisite for course y. Give an interpretation for each of the following:
   (a) Path
   (b) Chain
   (c) Cycle
   (d) Circuit
   (e) Connected component

2. The *inner degree* $d^-(x)$ of vertex x is defined as the number of arcs whose head is vertex x. The *outer degree* $d^+(x)$ of vertex x is defined as the number of arcs whose tail is vertex x. The *degree* $d(x)$ of vertex x is defined as the sum of its inner and outer degrees.

   Show that for any graph G the number of vertices with odd degree is even. Show that for any graph $G = (X, A)$

   $$\sum_{x \in X} d^+(x) = \sum_{x \in X} d^-(x)$$

3. Is it possible for a cut and a cycle to contain exactly one arc in common? Why?

4. Given a spanning tree T for graph G, show there exists a unique chain consisting exclusively of edges in T such that each edge joins any two vertices in G.

5. Consider the following linear programming problem:

   Maximize
   $$3x_1 + 2x_2 + 1x_3 - 4x_4$$
   such that
   $$2x_1 + 1x_2 + 3x_3 + 7x_4 \leq 10$$
   $$x_1 \geq 0, \quad x_2 \geq 0, \quad x_3 \geq 0, \quad x_4 \geq 0$$

   (a) Find an optimal solution. (Hint: Consider only basic solutions.)

(b) What is the dual linear programming problem?

(c) Use the complementary slackness conditions to find an optimal solution to the dual.

6. Is every subset of a tree also a tree? Is every subset of a forest also a forest?

7. In Fig. 1.3, do the arcs α, γ, ε form a chain? a path? Do the arcs α, γ, γ, ε form a chain? a path?

8. Suppose forest F consists of t trees and contains v vertices. How many arcs are in forest F?

9. Many of the Common Market countries share common borders. Construct a graph G whose vertices represent the Common Market countries. Join two vertices by an arc if the corresponding countries have a common border. Is G connected? Find the cut set with the smallest number of arcs. Is there any country whose withdrawal from the Common Market would sever all land travel between the remaining countries?

REFERENCES

Graph Theory

Berge, C., 1973. *Graphs and Hypergraphs* (translated by E. Minieka), North-Holland, Amsterdam.

Busacker, R., and T. Saaty, 1965. *Finite Graphs and Networks*, McGraw-Hill, New York.

Ford, L. R., and D. R. Fulkerson, 1962. *Flows in Networks*, Princeton Press, Princeton.

Frank, H., and I. Frisch, 1971. *Communication, Transmission, and Transportation Networks*, Addison-Wesley, Reading.

Harary, F., 1969. *Graph Theory*, Addison-Wesley, Reading.

Hu, T. C., 1969. *Integer Programming and Network Flows*, Addison-Wesley, Reading.

Ore, O., 1963. *Graphs and Their Uses, Random House New Mathematical Library*, Random House, New York.

Potts, R. B., and R. M. Oliver, 1972. *Flows in Transportation Networks*, Academic Press, New York.

Wilson, R., 1972. *Introduction to Graph Theory*, Academic Press, New York.

References

Linear Programming

Dantzig, G. B., 1963. *Linear Programming and Extensions*, Princeton Press, Princeton.

Hadley, G., 1962. *Linear Programming*, Addison-Wesley, Reading.

Simonnard, M., 1966. *Linear Programming*, (translated by W. S. Jewell), Prentice-Hall, Englewood.

Chapter 2

TREE ALGORITHMS

A graph may contain many different trees. This chapter studies several algorithms used to construct trees with certain optimal properties.

2.1 SPANNING TREE ALGORITHMS

Consider a graph $G = (X, E)$ in which the direction of each arc is unspecified. Suppose there is a weight $a(x,y)$ assigned to each edge (x,y) in graph G. Define the *weight of a tree* as the sum of the weights of the edges in the tree.

In this section, we shall first consider an algorithm to construct a spanning tree of graph G. Secondly, we shall consider an algorithm to construct a minimum weight spanning tree, i.e., a spanning tree of graph G whose weight is less than or equal to the weight of every other spanning tree of graph G.

EXAMPLE 1 (Rumor Monger). Consider a small village in which some of the villagers have a daily chat with one another. Is it possible for a rumor to pass throughout the entire village?

To answer this question, represent each villager by a vertex. Join two vertices by an edge if the corresponding two villagers have a daily chat with one another. If the resulting graph is connected, then it is possible for a rumor to pass through the entire village. To determine if the graph is connected, we could see whether the

19

graph possesses a spanning tree. If the graph possesses no spanning tree, then it cannot be connected and the rumor cannot circulate throughout the entire village.

EXAMPLE 2. The Department of Highways wishes to build enough new roads so that the five towns in a certain county will all be connected to one another either directly or via another town. The cost of constructing a highway between each pair of towns is known (see Fig. 2.1). Let each town correspond to a vertex, and let each possible highway to be constructed correspond to an edge joining

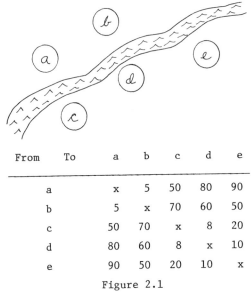

From	To	a	b	c	d	e
a		x	5	50	80	90
b		5	x	70	60	50
c		50	70	x	8	20
d		80	60	8	x	10
e		90	50	20	10	x

Figure 2.1

Highway Construction

the vertices that represent the two town joined by the highway. Associate a weight to each edge that is equal to the cost of constructing the corresponding highway. Deciding which highways to build can be viewed as the problem of constructing a minimum cost spanning tree for the corresponding graph. This follows since the edges of any spanning tree will connect each vertex (town) with every other vertex (town). Moreover, the minimum weight spanning tree will represent a minimum total cost set of new highways.

2.1 Spanning Tree Algorithms

Note that if the highways are allowed to have junctions at places other than the five towns, then the resulting problem is more complicated than a minimum spanning tree problem.

The *spanning tree algorithm*, given in the following discussion, is one of the most elegant algorithms that we shall encounter. The algorithm examines the edges in any arbitrary sequence and decides whether each edge will be included in the spanning tree. In the jargon of the algorithm, examining the edge is called *coloring* the edge. Blue is the color for edges included in the spanning tree; orange is the color for edges excluded from the spanning tree.

When an edge is examined, the algorithm simply checks if the edge under consideration forms a cycle with the other edges already assigned to the tree (blue edges). If so, then the edge under examination is excluded from the tree (colored orange); otherwise, this edge is assigned to the tree (colored blue).

How does the algorithm determine if the edge under consideration forms a cycle with the edges already assigned to the tree (blue edges)? As the edges are assigned to the tree they form one or more connected components. The vertices belonging to a single connected component are collected together into what the algorithm terms a "bucket". An edge forms a cycle with the edges already assigned to the tree if both its endpoints are in the same connected component (bucket).

The algorithm terminates with a spanning tree when the number-of-vertices-less-one edges have been colored blue, or equivalently, when all vertices are in one bucket, since all spanning trees must consist of the number-of-vertices-less-one edges. If the graph contains no spanning tree, which is equivalent to being not connected, then the algorithm terminates after coloring all edges without coloring enough blue edges.

Spanning Tree Algorithm

Initially, all edges are uncolored and all buckets are empty.

Step 1: Select any edge that is not a loop. Color this edge blue and place both its endpoints into an empty bucket.

Step 2: Select any uncolored edge that is not a loop. (If no such edge exists, stop the algorithm; no spanning tree exists.) One of four different situations must occur:

(a) Both endpoints of this edge are in the same bucket.
(b) One endpoint of this edge is in a bucket, the other endpoint is not in any bucket.
(c) Neither endpoint is in any bucket.
(d) Each endpoint is in a different bucket.

If item (a) occurs, color the edge orange (not in the tree) and return to Step 2. If (b) occurs, color the edge blue (in the tree) and assign the unbucketed endpoint to the same bucket as the other endpoint. If (c) occurs, color the edge blue and assign both endpoints to an empty bucket. If (d) occurs, color the edge blue and combine the contents of both buckets into one bucket, leaving the other bucket empty. Go to Step 3.

Step 3: If all the vertices of the graph are in one bucket, stop the algorithm since the blue edges form a spanning tree. Otherwise, return to Step 2.

Note that each time a step of the algorithm is performed an edge is indelibly colored. If there are only a finite number of edges in the graph, then the algorithm must stop after a finite number of steps.

If the algorithm does not terminate with a spanning tree, then no spanning tree exists for the graph for the following reason: The algorithm will terminate with two sets (buckets) of vertices that have no edge joining a member of one set to a member of the other set. Otherwise, such an edge would have been colored blue by the algorithm and the two buckets would have been merged together. Hence, the algorithm does what it is supposed to do, construct a spanning tree.

This algorithm has the property that each edge is examined at most one time. Once the algorithm colors an edge, this edge is never considered again. Algorithms, like the spanning tree algorithm,

2.1 Spanning Tree Algorithms

which examine each entity at most once and decide its fate once and for all during that examination are called *greedy* algorithms. The advantage to performing a greedy algorithm is that you do not have to spend your time reexamining entities, and it is easier to determine the maximum number of operations that will have to be performed by the algorithm. In the case of the spanning tree algorithm, one will never have to perform more steps than there are edges in the graph.

EXAMPLE 3. Let us construct a spanning tree for the graph in Fig. 2.2. Examine the edges in the following arbitrary order: (a,b), (d,e), (a,d), (b,e), (e,d), (c,b), (a,c), and (c,d). The results of each step of the algorithm are shown below:

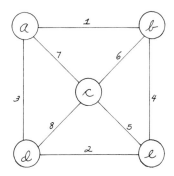

Figure 2.2

Edge	Color	Bucket No. 1	Bucket No. 2
		Initially empty	Initially empty
(a,b)	Blue	a, b	Empty
(d,e)	Blue	a, b	d, e
(a,d)	Blue	a, b, d, e	Empty
(b,e)	Orange	a, b, d, e,	Empty
(e,c)	Blue	a, b, d, e, c	Empty

Stop since all vertices are in one bucket. The four blue edges (a,b), (d,e), (a,d), (e,c) form a spanning tree of the graph.

Obviously, the spanning tree constructed by the algorithm depends upon the order in which the edges are examined by the algorithm. If the edges in Example 3 had been examined in the reverse order, then the algorithm would have generated the spanning tree consisting of edges (c,d), (a,c), (c,b), (e,d).

Now consider the problem of finding a spanning tree with the smallest possible weight or the largest possible weight, respectively called a *minimum spanning tree* and a *maximum spanning tree*. Obviously, if a graph possesses a spanning tree, it must have a minimum spanning tree and also a maximum spanning tree. These spanning trees can be readily constructed by performing the spanning tree algorithm with an appropriate ordering of the edges.

Minimum Spanning Tree Algorithm

Perform the spanning tree algorithm examining the edges in order of ascending weight (smallest first, largest last). If two or more edges have the same weight, order them arbitrarily.

Maximum Spanning Tree Algorithm

Perform the spanning tree algorithm examining the edges in order of descending weight (largest first, smallest last). If two or more edges have the same weight, order them arbitrarily.

Proof of the Minimum Spanning Tree Algorithm: We shall prove by contradiction that the minimum spanning tree algorithm constructs a minimum spanning tree. Suppose that the algorithm constructs a tree T and some other tree S is in fact a minimum spanning tree. Since S and T are not identical trees, they differ by at least one edge. Denote by $e_1 = (x,y)$ the first examined edge that is in T but not in S. Since S is a spanning tree, there exists in S a unique chain from vertex x to vertex y. Call this chain $C(x,y)$. If edge $e_1 = (x,y)$ is added to tree S, then a cycle is formed by e_1 and $C(x,y)$. Since tree T contains no cycles, this cycle must contain at least one edge e_2 not contained in tree T.

2.1 Spanning Tree Algorithms

Remove edge e_2 from S and add edge e_1 to S. Call the resulting set of edges S'. Clearly, S' is also a spanning tree. Since S is by definition a minimum spanning tree, the weight of S' must be greater than or equal to the weight of S, and hence

$$a(e_1) \geq a(e_2).$$

Suppose that edge e_2 was examined before edge e_1 in the minimum spanning tree algorithm that generated tree T. Since e_2 was not included in tree T, edge e_2 must form a cycle with the edges of tree T that were examined before edge e_2. But since e_1 is defined as the first edge in T that is not in S, all other edges in this cycle must be in tree S. This is a contradiction since edge e_2 forms a cycle with these edges. Therefore, we must conclude that edge e_1 was examined before edge e_2, and $a(e_2) \geq a(e_1)$. Therefore, trees S and S' have the same total cost, and tree S' has one more edge in common with tree T than does spanning tree S.

The proof can now be repeated using S' as the minimum cost spanning tree instead of S. This generates another minimum spanning tree S'' that has one more edge in common with tree T than did tree S'. Ultimately, a minimum spanning tree will be generated that is identical to tree T. Thus, tree T is a minimum spanning tree. Q.E.D.

The proof for the maximum spanning tree algorithm is identical to the preceding proof except that minimum should everywhere be replaced by maximum.

Construct a minimum cost spanning tree for the highway problem in Example 2. As shown in Fig. 2.1, the edges in ascending order of cost are (a,b), (c,d), (d,e), (c,e), (a,c), (b,e), (b,d), (b,c), (a,d), and (a,e). The results of the minimum spanning tree algorithm are:

Edge	Color	Bucket no. 1	Bucket no. 2
		(Initially empty)	(Initially empty)
(a,b)	Blue	a, b	Empty
(c,d)	Blue	a, b	c, d
(d,e)	Blue	a, b	c, d, e
(c,e)	Orange	a, b	c, d, e
(a,c)	Blue	a, b, c, d, e	

Stop, all vertices are in the same bucket and four edges have been colored blue. A minimum cost spanning tree is (a,b), (c,d), (d,e), (a,c). What is the total cost?

The tree (a,b), (c,d), (d,e), (b,e) is also a minimum cost spanning tree.

2.2 MAXIMUM BRANCHING ALGORITHM

In Section 2.1, we considered tree-generating algorithms that ignore the direction of the arcs. In this section, we shall study an algorithm, due to Edmonds (1968), called the *maximum branching algorithm* that considers the direction of each arc.

Suppose the sales manager of a multinational company has a message that he wants conveyed to each of his district managers. What is the best way for him to accomplish this? One solution might be for him to phone each district manager personally. However, this might be very costly for the following reason: Suppose the sales manager is in Chicago and there are district managers in London and Paris. It would be far more expensive for him to phone each of them personally than it would be for him to phone London and have the London manager phone Paris.

Perhaps a better solution would be to have each district manager know in advance the selected district managers to whom he is to pass the message after he receives it, i.e., a "grapevine."

What properties should such a grapevine have? Obviously, each person should receive the message exactly once, and the cost of operating the grapevine should be as small as possible.

Construct a graph in which each vertex corresponds to a person in the grapevine and each arc represents the way a message may be transmitted between two people.

An *arborescence* is defined as a tree in which no two arcs are directed into the same vertex (see Fig. 2.3). Note that several arcs in an arborescence can share a common tail vertex. An arborescence can be thought of as a directed tree that can be used as a grapevine. The *root* of an arborescence is the unique vertex included

2.2 Maximum Branching Algorithm

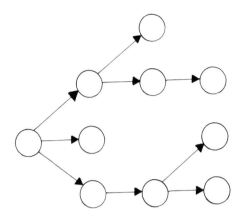

Figure 2.3
An Arborescence

in the arborescence that has no arcs directed into it. A *branching* is defined as a forest in which each tree is an arborescence.

A *spanning arborescence* is an arborescence that is also a spanning tree. A *spanning branching* is any branching that includes every vertex in the graph.

Associate a *weight* a(x,y) to each arc (x,y). The *weight of an arborescence* (or of a *branching*) is defined as the sum of the weights of the arcs in the arborescence (or in the branching).

A *maximum branching of graph G* is any branching of graph G with the largest possible weight. A *maximum arborescence of graph G* is any arborescence of graph G with the largest possible weight. A *minimum branching* and a *minimum arborescence* are defined similarly.

This section presents an algorithm due to Edmonds (1968) called the *maximum branching algorithm*. This algorithm, as its name suggests, constructs a maximum branching for any graph G.

As shown in the following, the maximum branching algorithm can also be used to find (1) a minimum branching, (2) a maximum spanning arborescence (if one exists), (3) a minimum spanning arborescence (if one exists), (4) a maximum spanning arborescence rooted at a specified vertex (if one exists), and (5) a minimum spanning arborescence rooted at a specified vertex (if one exists).

1. A *minimum branching* is found by replacing each arc's weight by its negative. A maximum branching for the new arc weights corresponds to a minimum branching for the original arc weights.
2. The algorithm finds a *maximum spanning arborescence* as follows: Suppose a positive constant M is added to each arc's weight. Obviously, as M increases, a maximum branching for the graph will contain more and more arcs. Since no branching can contain more arcs than a spanning arborescence, the maximum branching produced by the algorithm will be a spanning arborescence (if one exists) for M large enough. Moreover, this spanning arborescence will be a maximum spanning arborescence.

Note that even if all arc weights are positive and the graph possesses a spanning arborescence, a maximum branching need not be a spanning arborescence (see, e.g., Fig. 2.4).

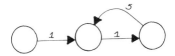

Figure 2.4

(Maximum branching has weight 5; spanning arborescence has weight 2.)

3. A *minimum spanning arborescence* (if one exists) can also be generated by the algorithm. This is accomplished by replacing each arc's weight by its negative. Then, a large positive constant M should be added to each arc's weight. As mentioned in item 2, the effect of adding M to each arc's weight is to force the maximum branching algorithm to generate only branchings with the maximum possible number of arcs. Hence, the algorithm will generate a spanning arborescence (if one exists). Moreover, this spanning arborescence will be a minimum spanning arborescence.
4. and 5. The maximum branching algorithm can also be used to find a *maximum* (or *minimum*) *spanning arborescence rooted at a specified vertex*, say vertex a. (The sales manager mentioned at the beginning of this section is seeking a minimum weight

2.2 Maximum Branching Algorithm

spanning arborescence rooted at Chicago.) This is accomplished by appending to the graph an additional vertex a' and an arc (a', a) with arbitrary weight. If the appended graph possesses a spanning arborescence, then this arborescence must be rooted at vertex a' since no arcs enter vertex a'. Any spanning arborescence rooted at a' must correspond to a spanning arborexcence rooted at a in the original graph. Moreover, this spanning arborescence rooted at vertex a. As described in items 2. and 3. the arc weights can be altered so that the maximum branching algorithm will be forced to find a maximum (or minimum) spanning arborescence, if one exists.

As described in items 2. and 3. the arc weights can be altered so that the maximum branching algorithm will be forced to find a maximum (or minimum) spanning arborescence, if one exists.

We shall now proceed to describe the *maximum branching algorithm*.

The maximum branching algorithm uses two buckets, the vertex bucket and the arc bucket. The vertex bucket contains only vertices that have been examined; the arc bucket contains arcs tentatively selected for the maximum branching. At all times, the arcs in the arc bucket form a branching. Initially both buckets are empty.

The algorithm successively examines the vertices in any arbitrary order. The examination of a vertex consists entirely of selecting the arc with the greatest positive weight that is directed into the vertex under examination (if any). If the addition of this arc to the arcs already selected for the arc bucket maintains a branching, then this arc is added to the arc bucket. Otherwise, this arc would form a *circuit* with some arcs already in the arc bucket. If this happens, then a new, smaller graph is generated by "shrinking" the arcs and vertices in this circuit into a single vertex, (see Fig. 2.5). Some of the arc costs are judiciously altered in the new, smaller graph. The vertex and arc buckets are redefined for the new graph as containing only their previous contents that appear in the new graph. The examination of each vertex continues as before. The process stops when all vertices have been examined.

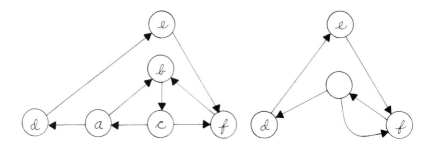

Figure 2.5
Before and After Shrinking Circuit
(a,b), (b,c), (c,a)

Upon termination, the arc bucket contains a branching for the final graph. The final graph is expanded back to its predecessor by expanding out its "artificial" vertex into a circuit. All but one of the arcs in this circuit are added to the arc bucket. The arc that is not added to the arc bucket is carefully selected so that the contents of the arc bucket remain a branching. This process is repeated until the original graph is regenerated. The arcs in the arc bucket upon termination turn out to be a maximum branching.

Denote the original graph for which the maximum branching is sought by G_0, and denote each successive graph generated from G_0 by G_1, G_2, ... The vertex and arc buckets used for these graphs will be denoted by V_0, V_1, ..., and A_0, A_1, ..., respectively. We are now ready to state formally the algorithm.

Maximum Branching Algorithm

Initially, all buckets V_0, V_1, ..., and A_0, A_1, ..., are empty. Set i = 0.

Step 1: If all vertices of G_i are in bucket V_i, go to Step 3. Otherwise, select any vertex v in G_i that is not in bucket V_i. Place vertex v into bucket V_i. Select an arc γ with the greatest positive weight that is directed into v. If no such arc exists, repeat Step 1; otherwise, place arc γ into bucket A_i. If the arcs in A_i still form a branching repeat Step 1; otherwise, go to Step 2.

2.2 Maximum Branching Algorithm

Step 2: Since the addition of arc γ to A_i no longer causes A_i to form a branching, arc γ forms a circuit with some of the arcs in A_i. Call this circuit C_i. Shrink all the arcs and vertices in C_i into a single vertex called v_i. Call this new graph G_{i+1}. Thus, any arc in G_i that was incident to exactly one vertex in C_i will be incident to vertex v_i in graph G_{i+1}. The vertices of G_{i+1} are v_i and all the vertices of G_i not in C_i.

Let the weights of each arc in G_{i+1} be the same as its weight in G_i except for the arcs in G_{i+1} that are directed into v_i. For each arc (x,y) in G_i that transforms into an arc (x,v_i) in G_{i+1}, let

$$a(x,v_i) = a(x,y) + a(r,s) - a(t,y) \qquad (1)$$

where (r,s) is the minimum weight arc in circuit C_i, and where (t,y) is the unique arc in circuit C_i whose head is vertex y. [At this point, observe that

$a(r,s) \geq 0$

$a(t,y) \geq a(r,s)$

and

$a(t,y) \geq a(x,y)$

since arc (t,y) was selected as the arc directed into vertex y.]

Let V_{i+1} contain all the vertices in G_{i+1} that are in V_i. (Thus, $v_i \in V_{i+1}$.) Let A_{i+1} contain all the arcs in G_{i+1} that are in A_i. (Thus, A_{i+1} contains the arcs in A_i that are not in C_i.)

Increase i by one, and return to Step 1.

Step 3: This step is reached only when all vertices of G_i are in V_i and the arcs in A_i form a branching for G_i. If $i = 0$, stop since the arcs in A_0 form a maximum branching for G_0. If $i \neq 0$, two cases are possible:

(a) Vertex v_{i-1} is the root of some arborescence in branching A_i.
(b) Vertex v_{i-1} is not the root of some arborescence in branching A_i.

If (a) occurs, then consider the arcs in A_i together with the arcs in circuit C_{i-1}. These arcs contain exactly one circuit in graph G_{i-1}, namely C_{i-1}. Delete from this set of arcs the arc in C_{i-1} that has the smallest weight. The resulting set of arcs forms a branching for graph G_{i-1}. Redefine A_{i-1} to be this set of arcs.

If (b) occurs, then there is a unique arc (x, v_{i-1}) in A_i that is directed into vertex v_{i-1}. This arc (x, v_{i-1}) corresponds in graph G_{i-1} to another arc, say arc (x,y), where vertex y is one of the vertices in circuit C_{i-1} that was shrunk to form vertex v_{i-1}. Consider the set of arcs in A_i together with the arcs in circuit C_{i-1}. This set of arcs contains exactly one circuit in G_{i-1}, namely C_{i-1}, and exactly two arcs directed into vertex y, namely arc (x,y) and an arc in circuit C_{i-1}. Delete this latter arc from this set of arcs. The remaining arcs in this set form a branching in graph G_{i-1}. Redefine A_{i-1} to be this set of arcs.

Having redefined A_{i-1}, decrease i by one unit and repeat Step 3.

Proof of the Maximum Branching Algorithm: Consider any graph G_t produced by the algorithm and consider the branching A_t produced by Step 3 for graph G_t. First, it will be shown that if A_t is a maximum branching for graph G_t, then branching A_{t-1} is a maximum branching for graph G_{t-1}.

To prove this, some definitions are needed. Let G' denote the subgraph consisting of all arcs in G_{t-1} not directed into a vertex in circuit C_{t-1}. Let G" denote the subgraph consisting of all the arcs in G_{t-1} not in G'. Thus, every arc of G_{t-1} is present in exactly one of these subgraphs G' and G". Let A'_{t-1} denote the arcs in A_{t-1} that are in G', and let A''_{t-1} denote the arcs of A_{t-1} that are in G". Clearly, A'_{t-1} and A''_{t-1} are branchings in G' and G", respectively.

If branching A_{t-1} is not a maximum branching for graph G_{t-1}, then there exists some branching B with greater total weight. Let B' denote the arcs in B that are in G', and let B" denote the arcs of B that are in G". Since B is a maximum branching, then it follows that either

      B' weighs more than A'_{t-1}

or

      B" weighs more than A''_{t-1}.

2.2 Maximum Branching Algorithm

Claim 1: A'_{t-1} is a maximum weight branching for G'.

Claim 2: A''_{t-1} weighs as much as B''.

If both Claims 1 and 2 are true, then it follows that A_{t-1} must be a maximum branching for graph G_{t-1}.

Note that the branching A_i produced by the algorithm for the terminal graph G_i is a maximum branching since it contains a maximum positively weighted arc directed into each vertex in G_i if such an arc exists. Since the algorithm produces a maximum branching for the terminal graph G_i, then if both Claims 1 and 2 are true, the algorithm must produce a maximum branching A_{i-1} for graph G_{i-1}. By repeating this reasoning, we can conclude that if Claims 1 and 2 are true then the branching A_0 produced by the algorithm is a maximum branching for the original graph G_0.

Hence, it remains only to show that Claims 1 and 2 are valid.

Proof of Claim 1: Suppose that circuit C_{t-1} contains n vertices. There is one arc with positive weight directed into each of these n vertices in graph G'. (Otherwise, the algorithm would not have formed circuit C_{t-1}.) Since there are only n vertices in G' that have arcs directed into themselves, a maximum branching for G' cannot contain more than n arcs. Moreover, no branching in G' can have weight exceeding the weight of circuit C_{t-1}, which consists of the maximum positive-weight arc directed into each of the n vertices in circuit C_{t-1}. However, at least one of the arcs in C_{t-1} must be absent from any maximum branching for G' since a branching cannot contain a circuit. Thus, at least one of these n vertices, say vertex $y \in C_{t-1}$, must either have no branching arc directed into it or else have an arc (x,y), $x \notin C_{t-1}$, directed into it.

For each vertex $z \in C_{t-1}$, construct a branching B_z in G' as follows:

(a) Include all arcs in circuit C_{t-1} except the arc in circuit C_{t-1} that is directed into vertex z
(b) Include any maximum positive-weight arc (x,z), where $x \notin C_{t-1}$.

Select the branching B_z^* with the greatest weight. From equation (1), branching B_z^* is the branching A'_{t-1} generated by the algorithm.

Consider any branching B_1 in G' that is not of the form B_z. If only one of the arcs of C_{t-1} is not in B_1, then it follows that B_1 cannot be a maximum branching for G' since it is not of the form B_z.

If two or more arcs of C_{t-1} are not in B_1, then each of these arcs is either (1) replaced by an arc of smaller weight directed into the same vertex or (2) no arc is directed into this vertex. In either case, this results in the decrease of the weight of arcs in the branching directed into the vertex. Hence, B_1 cannot be a maximum branching for G'.

Thus, A'_{t-1} is a maximum branching for G', and we can assume, without loss of generality, that A'_{t-1} is identical to B'. This concludes the proof of Claim 1.

Proof of Claim 2: Two cases are possible:

(a) Branching A_t contains an arc (x, v_{t-1}) directed into v_{t-1}.
(b) Branching A_t does not contain an arc directed into vertex v_{t-1}.

Case (a): By hypothesis, A_t is a maximum branching for G_t and contains an arc (x, v_{t-1}) directed into v_{t-1}. From Claim 1, B' is identical to A'_{t-1} and hence B' contains an arc (x,y), where $x \in C_{t-1}$ and $y \in C_{t-1}$. Since B is a branching in G_{t-1}, it follows that B'' cannot contain a path of arcs from a vertex in C_{t-1} to vertex x. Thus, B'' must be a maximum branching for G'' that does not contain a path of arcs from a vertex in C_{t-1} to vertex x.

Each arc in G'' corresponds to an arc with identical weight in G_t. Moreover, each branching in G'' corresponds to a branching in G_t with identical weight. Consequently, if A''_{t-1} is not a maximum branching in G'' that contains no path of arcs from a vertex in C_{t-1} to a vertex x, then A_t is not a maximum branching in G_t that contains arc (x, v_{t-1}), which is impossible. Hence, A''_{t-1} has the same weight as B'', which proves the claim for case (a).

Case (b): Each arc in G'' corresponds to an arc with identical weight in G_t. By hypothesis A_t is a maximum branching for G_t. Since no arc in A_t is directed into v_{t-1}, it follows that every

2.2 Maximum Branching Algorithm

arc in A_t corresponds to an arc in A''_{t-1}. Moreover, any branching in G'' corresponds to a branching in G_t with the same weight. Hence if A''_{t-1} were not a maximum branching in G'', then A_t would not be a maximum branching in G_t, which is a contradiction.

Thus, A''_{t-1} must be a maximum branching in G'' and have the same weight as B'', which completes the proof of Claim 2. Q.E.D.

EXAMPLE (Maximum Branching Algorithm). We shall now perform the maximum branching algorithm to find a maximum weight branching for the graph in Fig. 2.6. The weight of each arc is shown next to the arc. The algorithm will arbitrarily examine the vertices in alphabetical order. The result of the examination of the first four vertices a, b, c, d is shown in Fig. 2.6.

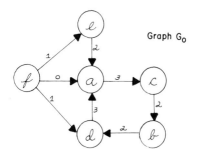

Graph G_0

Vertex examined	V_0	A_0
a	a	(d,a)
b	a,b	(d,a), (c,b)
c	a,b,c	(d,a), (c,b), (a,c)
d	a,b,c,d	(d,a), (c,b), (a,c), (b,d)

Figure 2.6

Maximum Branching Algorithm

After vertex d has been examined, the arcs in bucket A_0 no longer form a branching since they contain a cycle (a,c), (c,b), (b.d), (d,a). At this point, the algorithm shrinks this cycle into a vertex v_0. Figure 2.7 displays the new graph G_1 resulting from this shrinking. The calculations of the weights are shown next to

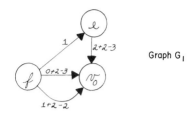

Graph G_1

Vertex examined	V_1	A_1
e	e	(f,e)
f	e,f	(f,e)
v_0	e,f,v_0	(f,e), (e,v_0)

Figure 2.7
Maximum Branching Algorithm

each arc in graph G_1. Graph G_1 has only three vertices e, f, and v_0. The result of the examination of each of these vertices is shown in Fig. 2.7.

After examining the three vertices in graph G_1, the algorithm has generated a maximum branching for graph G_1 consisting of arcs (f,e) and (e,v_0). Using this branching, Step 3 expands vertex v_0 back into its original cycle and adds arcs (a,c), (c,b), (b,d), and (d,a) to the arcs (f,e) and (e,v_0) already in the branching. Next, arc (d,a) is deleted from the branching so that only one branching arc, namely (e,a), is directed into vertex a. The resulting branching in graph G_0 consists of arcs (f,e), (e,a), (a,c), (c,b), and (b,d). The total weight of this branching equals $1 + 2 + 3 + 2 + 2 = 10$, which is the maximum possible weight.

Note that this branching also happens to be a spanning arborescence of graph G_0 rooted at vertex f.

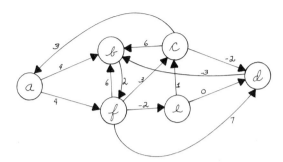

Figure 2.8

EXERCISES

1. Construct a minimum cost highway system for the five towns in Example 2.
2. Construct a minimum weight spanning tree for the graph in Fig. 2.8.
3. Construct a minimum weight spanning tree for the graph in Fig. 2.8 that includes arcs (a,b) and (c,d).
4. Construct a maximum weight branching for the graph in Fig. 2.8.
5. Construct a maximum spanning arborescence for the graph in Fig. 2.8 that is rooted at vertex a.
6. Suppose that you have just completed the maximum branching algorithm and discover that all your arc weights were understated by 5 units. Can you salvage your results, or is it necessary to repeat the algorithm again? Explain.
7. Consider the following greedy maximum branching algorithm:

    "Order the acrs according to their weights with the heaviest arc first. Select the first arc for the branching. Sequentially examine the rest of the arcs, selecting an arc for the branching if it forms a branching with the other arcs already selected for the branching."

Show that this algorithm does not always terminate with a maximum branching.

8. Restate the maximum branching algorithm for the special case when all arcs have the same weight.

9. A pipeline system must be built to connect the seven company-owned refineries with the port facility that receives imported crude oil. The cost of building the pipeline between any two points is $1000 per mile plus a $4000 set-up cost for each segment. The distance between all pairs of points is given in the following table. Find the least cost pipeline.

	P	R1	R2	R3	R4	R5	R6	R7
Port	0	5	6	8	2	6	9	10
Refinery 1		0	4	10	5	8	6	10
2			0	11	8	4	9	10
3				0	10	3	6	7
4					0	2	5	9
5						0	10	5
6							0	8
7								0

10. Suppose that the values in Exercise 9 represented the cost of shipping a year's supply of oil. Find the pipeline that would minimize the total yearly cost of shipments from the port to each refinery.

11. Each day, every researcher in a laboratory receives his assignment from one of his superiors. The researchers fall into four categories: senior, associate, assistant, and junior. There are, respectively, 1, 4, 6, 5, 8 persons currently in each category. Because of differences in sophistication and education, the time required to convey an assignment is as follows:

Exercises

From To	Associate	Assistant	Junior
Senior	5	9	11
Associate		4	8
Assistant			4

What is the best way to disseminate the daily assignments?

REFERENCES

Edmonds, J., 1968. Optimum Branchings, *Mathematics of the Decision Sciences, Lectures in Applied Mathematics*, Vol. 2, AMS, 1968 (G. Dantzig and A. Veinott, eds.), pp. 346-361.

Kruskal, J. B., 1956. On the Shortest Spanning Subtree of a Graph and the Traveling Salesman Problem, *Proc. AMS*, 7, pp. 48-50.

Rosentiehl, P., 1966. L'Arbre Minimum d'un Graphe, *International Seminar on Graph Theory*, Rome, July.

Chapter 3

PATH ALGORITHMS

This chapter describes several algorithms for finding paths with certain optimal properties. The first section presents an algorithm for finding the "shortest" path between two given vertices in a graph. Section 3.2 presents algorithms for finding the "shortest" path between every pair of vertices in the graph. Section 3.3 presents algorithms for finding the second, third. etc., shortest versions of the paths found above. Section 3.4 reconsiders these algorithms in a more general, algebraic framework.

3.1 SHORTEST PATH ALGORITHM

Associate a number $a(x,y)$ with each arc (x,y) in graph G. If no arc (x,y) exists in graph G, let $a(x,y) = \infty$. We shall refer to $a(x,y)$ as the *length* of arc (x,y), although $a(x,y)$ could also be regarded as the cost or the weight of arc (x,y). Let the *length of a path* be defined as the sum of the lengths of the individual arcs comprising the path.

For any two vertices s and t in graph G, it is possible that there exist several paths from s to t in graph G. In this section, we shall consider an algorithm that generates a path from s to t that has the smallest possible length. Such a path is called a *shortest path* from s to t.

EXAMPLE 1. Suppose that you wish to drive from Boston to Los Angeles using only interstate highways. What is the shortest route to take?

Construct a graph whose vertices correspond to the junctions of the interstate highways. Let the arcs correspond to the interstate highways joining their respective vertices. Let the length of each arc equal the mileage along the highway that it represents.

The routing problem between Boston and Los Angeles can now be solved by finding a shortest path from the vertex representing the highway junction in Boston where you would start your journey to the vertex representing the highway junction in Los Angeles where you would end your journey.

EXAMPLE 2. An airline is approached by a passenger who wishes to fly from Springfield, Illinois to Ankara, Turkey spending as little time as possible in the air since he is afraid of flying. How should the airline route this passenger?

The airline should construct a graph whose vertices are the airports between Springfield and Ankara and whose arcs correspodd to the flights between the corresponding airports. Let the length of each arc equal the corresponding flight time. The airline should now read the remainder of this section to learn how to find a shortest path between the vertices corresponding to Springfield and Ankara.

EXAMPLE 3. You must have an automobile at your disposal for the next five years until your retirement. There are currently a variety of automobiles that you may purchase with different expected lifetimes and costs and there are a variety of leasing arrangements available. What should you do?

Represent each possible transaction date during the next five years as a vertex. (To simplify, you might consider only the first day of each month as a possible transaction date.) Represent each possible purchase or lease by an arc from its initial transaction date vertex to its maturity date vertex. Let the length of each arc equal the cost of the corresponding transaction. The best purchase/lease combination must correspond to the shortest path from the vertex representing the current time to the vertex representing your retirement date.

3.1 Shortest Path Algorithm

EXAMPLE 4. Suppose that a salesman in Boston plans to drive to Los Angeles to visit an important client. He plans to use the interstate highway system described in Example 1, and he plans to visit other clients along his route. He knows on the average how much commission he can earn from each proposed stop along the way from Boston to Los Angeles. What route should he take from Boston to Los Angeles?

In this situation, the length of each arc in the graph representing the highway system should be set equal to the expected net cost (driving expenses less expected commission) of the corresponding highway segment. The salesmen should drive along the shortest path from the Boston vertex to the Los Angeles vertex.

Observe that in this situation arc costs will be negative on any arc where the salesman expects to make a profit and positive on any arc where the salesman expects to incur a loss. In Example 1, all arc costs were nonnegative. As we shall see later, these two situations require different solution algorithms.

EXAMPLE 5. A small investor must decide how to invest optimally his funds for the coming year. A variety of investments (passbook savings account, certificate of deposit, savings bonds, etc.) are available. For simplicity, suppose that investments can only be made and withdrawn on the first of each month. Create a vertex corresponding to the first day of each month. Place an arc from vertex x to vertex y for each investment that can be made at time x and mature at time y. Notice that this graph contains no circuits. Let the length of each arc equal the negative of the profit earned on the corresponding investment. The best investment plan corresponds to a shortest path (i.e., path with the most negative total length) from the vertex for time zero to the vertex for one year from now. This example is similar to Example 3 except that now arc lengths may be negative.

The algorithm presented here finds a shortest path between two specified vertices s and t when all arc lengths are nonnegative. This algorithm due to Dijkstra (1959) is generally acknowledged to be one of the most efficient algorithms for solving this problem.

The main idea underlying the Dijkstra *shortest path algorithm* is quite simple: Suppose we know the m vertices that are closest in total length to vertex s in the graph and also a shortest path from s to each of these vertices.[+] Color vertex s and these m vertices. Then, the m + 1-st closest vertex to s is found as follows:

For each uncolored vertex y, construct n distinct paths from s to y by joining the shortest path from s to x with arc (x,y) for all colored vertices x. Select the shortest of these m paths and let it tentatively be the shortest path from s to y.

Which uncolored vertex is the m + 1-st closest vertex to s? It is the uncolored vertex with the shortest tentative path from s as calculated above. This follows because the shortest path from s to the m + 1-st closest vertex to s must use only colored vertices as its intermediate vertices since all arc lengths are nonnegative.

So, if the m closest vertices to s are known, the m + 1-st can be determined as above. Starting with m = 0, this process can be repeated until the shortest path from s to t has been found.

With this in mind as motivation, we can now formally state the Dijkstra shortest path algorithm.

Dijkstra Shortest Path Algorithm

Step 1: Initially, all arcs and vertices are uncolored. Assign a number d(x) to each vertex x to denote the length of the shortest path from s to x that uses only colored vertices as intermediate vertices. Initially, set d(s) = 0 and d(x) = ∞ for all x ≠ s. Let y denote the last vertex to be colored.

Color vertex s and let y = s.

Step 2. For each uncolored vertex x, redefine d(x) as follows:

$$d(x) = \min\{d(x), d(y) + a(y,x)\} \tag{1}$$

If d(x) = ∞ for all uncolored x, then stop because no path exists from s to any uncolored vertex. Otherwise, color the uncolored vertex x with the smallest value of d(x). Also color the arc

---

[+]Of course, a shortest path from vertex s to itself is the null path (path with no arcs) which has length equal to zero.

3.1 Shortest Path Algorithm

directed into vertex x from a colored vertex that determined the value of d(x) in the above minimization. Let y = x.

Step 3. If vertex t has been colored stop because a shortest path from s to t has been discovered. This path consists of the unique path of colored arcs from s to t. If vertex t has not been colored yet, repeat Step 2.

Note that whenever the algorithm colors a vertex (except vertex s) the algorithm also colors an arc directed into this vertex. Thus, each vertex has at most one colored arc directed into it, and the colored arcs cannot contain a cycle since no arc is colored if both its endpoints have a colored arc incident to it. Therefore, we can conclude that the colored arcs form an arborescence rooted at s. This arborescence is called a *shortest path arborescence*. The unique path from s to any other vertex x contained in any shortest path arborescence is a shortest path from s to x.

If the shortest path from s to x in a shortest path arborescence passes through vertex y, then it follows that the portion of this path from y to x is a shortest path from y to x. Otherwise, there exists another, even shorter, path from y to x which contradicts that we found a shortest path from s to x.

Since the colored arcs at all times form an arborescence, the algorithm can be regarded as the growing of an arborescence rooted at vertex s. Once vertex t is reached, the growing process can be terminated.

If you wanted to determine a shortest path from vertex s to every other vertex in the graph, then the growing process could be continued until all vertices were included in the shortest path arborescence, in which case the arborescence would become a spanning arborescence (if one exists). In this case, Step 3 would read as follows:

If all vertices have been colored, stop because the unique path of colored arcs from s to x is a shortest path from s to x for all vertices x. Otherwise, return to Step 2.

EXAMPLE 6. Let us perform the Dijkstra shortest path algorithm to find a shortest path from vertex s to vertex t in the graph in Fig. 3.1.

Initially only vertex s is colored, d(s) = 0, and d(x) = ∞ for all x ≠ s.

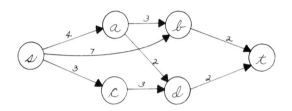

Figure 3.1

Step 2: y = s

d(a) = min{d(a), d(s) + a(s,a)} = min{∞, 0 + 4} = 4

d(b) = min{d(b), d(s) + a(s,b)} = min{∞, 0 + 7} = 7

d(c) = min{d(c), d(s) + a(s,c)} = min{∞, 0 + 3} = 3

d(d) = min{d(d), d(s) + a(s,d)} = min{∞, 0 + ∞} = ∞

d(t) = min{d(t), d(s) + a(s,t)} = min{∞, 0 + ∞} = ∞

Since d(c) = 3 = min{d(a), d(b), d(c), d(d), d(t)}, vertex c is colored and arc (s,c), which determined d(c), is colored. The current shortest path arborescence consists of arc (s,c) [see Fig. 3.2(a)].

Step 3: Vertex t has not been colored, return to Step 2.

Step 2: y = c

d(a) = min{d(a), d(c) + a(c,a)} = min{4, 3 + ∞} = 4

d(b) = min{d(b), d(c) + a(c,b)} = min{7, 3 + ∞} = 7

d(d) = min{d(d), d(c) + a(c,d)} = min{∞, 3 + 3} = 6

d(t) = min{d(t), d(c) + a(c,t)} = min{∞, 3 + ∞} = ∞

Since d(a) = 4 = min{d(a), d(b), d(d), d(t)}, vertex a is colored and arc (s,a), which determined d(a), is colored. The current shortest path arborescence consists of arcs (s,c) and (s,a) [see Fig. 3.2(b)].

3.1 Shortest Path Algorithm

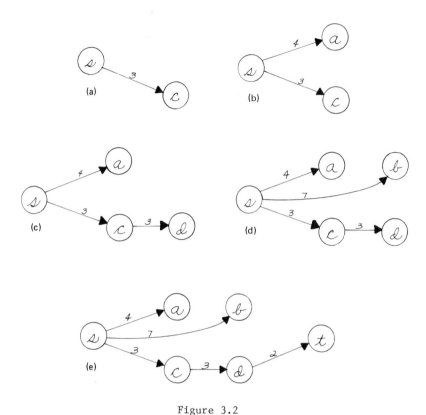

Figure 3.2
Growing a Shortest Path Arborescence

Step 3: Vertex t has not been colored, return to Step 2.

Step 2: y = a

$d(b) = \min\{d(b), d(a) + a(a,b)\} = \min\{7, 4 + 3\} = 7$
$d(b) = \min\{d(d), d(a) + a(a,d)\} = \min\{6, 4 + 2\} = 6$
$d(t) = \min\{d(t), d(a) + a(a,t)\} = \min\{\infty, 4 + \infty\} = \infty$

Since $d(d) = 6 = \min\{d(b), d(d), d(t)\}$, vertex d is colored. Both arcs (c,d) and (a,d) determined d(d), and we can arbitrarily select one of these two arcs to be colored. Let us arbitrarily select arc (c,d). Hence, the current shortest path arborescence becomes arcs (s,c), (s,a), and (c,d) [see Fig. 3.2(c)].

Step 3: Vertex t has not been colored, return to Step 2.

Step 2: y = d

$d(b) = \min\{d(b), d(d) + a(d,b)\} = \min\{7, 6 + \infty\} = 7$
$d(t) = \min\{d(t), d(d) + a(d,t)\} = \min\{\infty, 6 + 2\} = 8$

Since $d(b) = 7 = \min\{d(b), d(t)\}$, vertex b is colored and arc (s,b), which determined d(b), is colored. The current shortest path arborescence becomes arcs (s,c), (s,a), (c,d), and (s,b) [see Fig. 3.2(d)].

Step 3: Vertex t has not been colored, return to Step 2.

Step 2: y = b

$d(t) = \min\{d(t), d(b) + a(b,t)\} = \min\{8, 7 + 2\} = 8$

Thus, vertex t is colored at last. Also, arc (d,t), which determined d(t), is colored. The final shortest path arborescence consists of the arcs (s,c), (s,a), (c,d), (s,b), and (d,t) [see Fig. 3.2(e)].

A shortest path from s to t consists of arcs (s,c), (c,d), and (d,t) and has length $3 + 3 + 2 = 8$.

This path is not the only shortest path from s to t since the path (s,a), (a,d), (d,t) has length $4 + 2 + 2 = 8$. A shortest path from s to t will be unique if there is never any choice as to which arc to color.

Finally, note that if there were a tie for the vertex to be colored, i.e., two different uncolored vertices had the same minimum value of d(x), then the selection could be made arbitrarily. On the next iteration of Step 2, the other vertex would be colored.

Initially, we assumed that all arc lengths were nonnegative. What would happen in the Dijkstra shortest path algorithm if some of the arc lengths were negative? For example, consider the graph in Fig. 3.3. The shortest path from vertex s to vertex t is (s,a), (a,t), whose length is $+2 - 2 = 0$. The reader can easily verify that if the Dijkstra shortest path algorithm were applied to this graph, then the path (s,t) would erroneously be selected as the shortest path from vertex s to vertex t. Hence, there is no guarantee that the Dijkstra

3.1 Shortest Path Algorithm

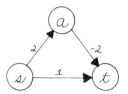

Figure 3.3

shortest path algorithm will produce a shortest path when arc lengths are permitted to be negative as in Examples 4 and 5.

Fortunately, the Dijkstra shortest path algorithm can be generalized to accomodate arcs with negative lengths. This generalization is due to Ford (1946). This is accomplished by three simple changes in the Dijkstra algorithm. They are:

1. In Step 2, let equation (1) be applied to all vertices, not only the uncolored vertices. Hence, a colored as well as an uncolored vertex can have its vertex number decreased.
2. If a colored vertex has its vertex number decreased, then uncolor the colored arc incident to it.
3. Terminate the algorithm only after all vertices are colored and Step 2 fails to lower any vertex numbers.

Proof: The proof of the Ford algorithm is achieved by contradiction. The Ford algorithm cannot terminate unless

$$d(x) + a(x,y) \leq d(y) \text{ [for all } (x,y)] \tag{2}$$

Otherwise, vertex y would be uncolored during the iteration of Step 2 occurring immediately after vertex x was colored.

Suppose that upon the termination of the algorithm d(y) does not equal the length of a shortest path from vertex s to vertex y for some vertex y. (If there is more than one such vertex y, then take y to be the vertex with the shortest path from s that contains the fewest arcs.) Whenever, a vertex number d(z) is finite, it equals the length of some path from vertex s to vertex z. Hence, it follows upon termination d(y) must equal the length of some path from s to y, and consequently, d(y) must exceed the length of a shortest path from s to y. Let vertex x be the next-to-last vertex on the shortest path from s to x. (If there is more than one such path, select any path with the fewest number of arcs.) Thus, d(x) must equal the length of a shortest path from s to x, and $d(y) > d(x) + a(x,y)$. This contradicts expression (2). Q.E.D.

EXAMPLE 7. Let us apply the Ford Algorithm to the graph in Fig. 3.3.

Step 1: Initially, only vertex s is colored, $d(s) = 0$, $d(a) = \infty$ and $d(t) = \infty$.

Step 2: y = s

$d(a) = \min\{d(a), d(s) + a(s,a)\} = \min\{\infty, 0 + 2\} = 2$
$d(t) = \min\{d(t), d(s) + a(s,t)\} = \min\{\infty, 0 + 1\} = 1$

Since $d(t) = \min\{d(a), d(t)\}$, vertex t is colored and arc (s,t) is also colored. The current shortest path arborescence is (s,t).

Step 3: Since not all vertices are colored, return to Step 2.

Step 2: y = t

Since there are no arcs leaving t, all vertex numbers remain unchanged. Hence, vertex a is colored and arc (s,a) is also colored. The shortest path arborescence now consists of arcs (s,t) and (s,a).

Step 3: Return to Step 2 to try to lower vertex numbers.

Step 2: y = a.

$d(t) = \min\{d(t), d(a) + a(a,t)\} = \min\{1, 2 - 2\} = 0$
$d(s) = \min\{d(s), d(a) + a(a,s)\} = \min\{0, 2 + \infty\} = 0$.

Since d(t) is reduced from 1 to 0, vertex t and arc (s,t) are uncolored. The shortest path arborescence now consists of only arc (s,a).

Vertex t is the only uncolored vertex, and hence by default, vertex t must be colored and arc (a,t) must also be colored. The shortest path arborescence is now (s,a) and (a,t).

3.1 Shortest Path Algorithm

Step 3: Return to Step 2 for y = t.

Step 2: y = t

Since there are no arcs leaving vertex t, no vertex numbers can be lowered. No vertices are uncolored.

Step 3: Since all vertices are colored and no vertex numbers can be lowered, the algorithm stops. The shortest path from s to t is (s,a), (a,t) whose length is 2 - 2 = 0.

Can the Ford algorithm fail? Yes, if the graph contains a circuit whose total length is negative (a negative circuit). In this case, the circuit might be repeated infinitely many times yielding a nonsimple path whose length is infinitely negative. For the salesman in Example 4, a negative circuit corresøonds to a profitable circular route that can be repeated infinitely many times yielding the salesman an infinite profit. In Example 5, the graph contained no circuits, so the Ford algorithm could be applied without hesitation.

What should we do if we are not certain if the graph under consideration contains any negative circuits? Apply the Ford algorithm anyway, but keep count on the number of times each vertex is colored. As soon as any vertex is colored for the N-th time, where N is the number of vertices in the graph, stop because the graph contains a negative circuit. Otherwise, the Ford algorithm terminates in a finite number of steps with correct results.

To show this, suppose that there are no negative circuits in the graph. When any vertex x receives its final vertex number value (i.e., the length of a shortest path from s to x), then at worst every other vertex can be colored once more before another vertex receives its final vertex number value. Hence, no vertex can be colored more than N - 1 times.

3.2 ALL SHORTEST PATH ALGORITHMS

The preceding section considered the problem of finding a shortest path from a specified vertex to every other vertex in the graph. In this section, we shall consider the problem of finding a shortest path between every pair of vertices in the graph. Of course, this second and more general problem could be solved by repeating the Dijkstra shortest path algorithm once for each vertex in the graph taken as the initial vertex s. However, this would require a great many computations, and fortunately algorithms exist that are more efficient than repeating the Dijkstra shortest path algorithm once for each vertex in the graph. This section presents two similar algorithms for finding all shortest paths. These algorithms are due to Floyd (1962) and Dantzig (1967). In both algorithms, arc lengths are permitted to be negative so long as no circuits have negative length.

EXAMPLE 1. The airline in Example 2 of Sec. 3.1 must daily route numerous passengers between the various cities in the United States. The airline for reasons of economy wants to route each passenger so that he minimizes his total mileage. Hence, the airline would like to have advance knowledge of the shortest air route between every pair of cities in the United States.

Before presenting the algorithms, some notation is needed. Number the vertices 1, 2, ..., N. Let d_{ij}^m denote the length of a shortest path from vertex i to vertex j, where only the first m vertices are allowed to be intermediate vertices. (Recall that an intermediate vertex is any vertex in the path, except the initial or terminal vertex of the path.) If no such path exists, then let $d_{ij}^m = \infty$. From this definition of d_{ij}^m it follows that d_{ij}^0 denotes the length of a shortest path from i to j that uses no intermediate vertices, i.e., the length of the shortest arc from i to j (if such an arc exists). Let $d_{ii}^0 = 0$ for all vertices i. Furthermore, d_{ij}^N represents the length of a shortest path from i to j.

Let D^m denote the N × N matrix whose i,j-th element is d_{ij}^m. If we know the length of each arc in the graph, then we can determine matrix D^0. Ultimately, we wish to determine D^N the matrix of shortest path lengths.

The Floyd shortest path algorithm starts with D^0 and calculates D^1 from D^0. Next, the Floyd shortest path algorithm calculates D^2 from D^1. This process is repeated until D^N is calculated from D^{N-1}.

3.2 All Shortest Path Algorithms

The basic idea underlying each of these calculations is the following: Suppose we know

(a) A shortest path from vertex i to vertex m that allows only the first m - 1 vertices as intermediate vertices.

(b) A shortest path from vertex m to vertex j that allows only the first m - 1 vertices as intermediate vertices.

(c) A shortest path from vertex i to vertex j that allows only the first m - 1 vertices as intermediate vertices. Since no circuits with negative length exist, then, the shorter of the paths given in items (d) and (e) must be a shortest path from i to j that allows only the first m vertices as intermediate vertices:

(d) The union of the paths in items (a) and (b)

(e) The path in item (c).

Thus,

$$d_{ij}^m = \min\{d_{im}^{m-1} + d_{mj}^{m-1}, d_{ij}^{m-1}\} \quad (3)$$

From equation (3), we can see that only the elements of matrix D^{m-1} are needed to calculate the elements of matrix D^m. Moreover, these calculations can be done without reference to the underlying graph. We are now ready to state formally the *Floyd shortest path algorithm* for finding a shortest path between each pair of vertices in a graph.

Floyd Shortest Path Algorithm

Step 1: Number the vertices of the graph 1, 2, ..., N. Determine the matrix D^0 whose i,j-th element equals the length of the shortest arc from vertex i to vertex j, if any. If no such arc exists, let $d_{ij}^0 = \infty$. Let $d_{ii}^0 = 0$ for i.

Step 2: For m = 1, 2, ..., N, successively determine the elements of D^m from the elements of D^m using the following recursive formula

$$d_{ij}^m = \min\{d_{im}^{m-1} + d_{mj}^{m-1}, d_{ij}^{m-1}\} \quad (3)$$

As each element is determined, record the path that it represents.

Upon termination, the i,j-th element of matrix D^N represents the length of a shortest path from vertex i to vertex j.

The optimality of this algorithm follows inductively from the fact that the length of a shortest path from i to j that allows only the first m vertices as intermediate vertices must be the smaller of (a) the length of a shortest path from i to j that allows only the first m - 1 vertices as intermediate vertices and (b) the length of a shortest path from i to j that allows only the first m vertices as intermediate vertices and uses the m-th vertex once as an intermediate vertex.

Note that $d_{ii}^m = 0$ for all i and for all m. Hence, the diagonal elements of the matrices $D^1, D^2, \ldots, D^N$ need not be calculated. Moreover, $d_{im}^{m-1} = d_{im}^m$ and $d_{mi}^{m-1} = d_{mi}^m$ for all i = 1, 2, ..., N. This follows since vertex m will not be an intermediate vertex in any shortest path starting or originating at vertex m since no circuits with negative length exist. Hence, in the computation of matrix D^m, the m-th row and the m-th column need not be calculated. Thus, in each matrix D^m only the (N - 1)(N - 2) elements that are neither on the diagonal nor in the m-th row or m-th column need be calculated.

EXAMPLE 2 (Floyd Shortest Path Algorithm). For the graph in Fig. 3.4, the matrix D^0 of arc lengths is

$$D^0 = \begin{bmatrix} 0 & 1 & 2 & 1 \\ 2 & 0 & 7 & \infty \\ 6 & 5 & 0 & 2 \\ 1 & \infty & 4 & 0 \end{bmatrix}$$

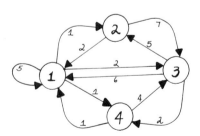

Figure 3.4

3.2 All Shortest Path Algorithms

The elements of D^1 and the corresponding shortest paths are calculated as follows:

$d_{ij}^1 = \min\{d_{i1}^0 + d_{1j}^0, d_{ij}^0\}$	Corresponding Path
$d_{11}^1 = d_{11}^0 = 0$	
$d_{12}^1 = d_{12}^0 = 1$	(1,2)
$d_{13}^1 = d_{13}^0 = 2$	(1,3)
$d_{14}^1 = d_{14}^0 = 1$	(1,4)
$d_{21}^1 = d_{21}^0 = 2$	(2,1)
$d_{22}^1 = 0$	
$d_{23}^1 = \min\{d_{21}^0 + d_{13}^0, d_{23}^0\} = \min\{2 + 2, 7\} = 4$	(2,1), (1,3)
$d_{24}^1 = \min\{d_{21}^0 + d_{14}^0, d_{24}^0\} = \min\{2 + 1, \infty\} = 3$	(2,1), (1,4)
$d_{31}^1 = d_{31}^0 = 6$	(3,1)
$d_{32}^1 = \min\{d_{31}^0 + d_{12}^0, d_{32}^0\} = \min\{6 + 1, 5\} = 5$	(3,2)
$d_{33}^1 = 0$	
$d_{34}^1 = \min\{d_{31}^0 + d_{14}^0, d_{34}^0\} = \min\{6 + 1, 2\} = 2$	(3,4)
$d_{41}^1 = d_{41}^0 = 1$	(4,1)
$d_{42}^1 = \min\{d_{41}^0 + d_{12}^0, d_{42}^0\} = \min\{1 + 1, \infty\} = 2$	(4,1), (1,2)
$d_{43}^1 = \min\{d_{41}^0 + d_{13}^0, d_{43}^0\} = \min\{1 + 2, 4\} = 3$	(4,1), (1,3)
$d_{44}^1 = 0$	

In a similar way, matrices D^2, D^3, and D^4 and the corresponding shortest paths can be calculated. The results of these calculations are

$$D^2 = \begin{bmatrix} 0 & 1 & 2 & 1 \\ 2 & 0 & 4 & 3 \\ 6 & 5 & 0 & 2 \\ 1 & 2 & 3 & 0 \end{bmatrix}$$

Shortest paths for D^2:

$$\begin{bmatrix} & (1,2) & (1,3) & (1,4) \\ (2,1) & & (2,1), (1,3) & (2,1), (1,4) \\ (3,1) & (3,2) & & (3,4) \\ (4,1) & (4,1), (1,2) & (4,1), (1,3) & \end{bmatrix}$$

$$D^3 = \begin{bmatrix} 0 & 1 & 2 & 1 \\ 2 & 0 & 4 & 3 \\ 6 & 5 & 0 & 2 \\ 1 & 2 & 3 & 0 \end{bmatrix}$$

Shortest paths for D^3:

$$\begin{bmatrix} & (1,2) & (1,3) & (1,4) \\ (2,1) & & (2,1), (1,3) & (2,1), (1,4) \\ (3,1) & (3,2), & & (3,4) \\ (4,1) & (4,1), (1,2) & (4,1), (1,3) & \end{bmatrix}$$

$$D^4 = \begin{bmatrix} 0 & 1 & 2 & 1 \\ 2 & 0 & 4 & 3 \\ 3 & 4 & 0 & 2 \\ 1 & 2 & 3 & 0 \end{bmatrix}$$

Shortest paths for D^4:

$$\begin{bmatrix} & (1,2) & (1,3) & (1,4) \\ (2,1) & & (2,1), (1,3) & (2,1), (1,4) \\ (3,4), (4,1) & (3,4), (4,1), (1,2) & & (3,4) \\ (4,1) & (4,1), (1,2) & (4,1), (1,3) & \end{bmatrix}$$

Note that since the numbering of the vertices is arbitrary, the algorithm will find the shortest paths at earlier iterations if

3.2 All Shortest Path Algorithms

vertices with numbers close to one another are in fact vertices that are "close" to one another.

In the preceding numerical example, the actual arcs that comprise each shortest path were recorded as the Floyd algorithm was performed. For problems of any realistic size, it is obvious that this procedure would be impractical. Consequently, we need to develop a more efficient technique for determining the actual arcs that constitute the shortest path.

The next-to-last vertex in a path is called the *penultimate* vertex of that path. Let p_{ij} denote the penultimate vertex of the shortest path from vertex i to vertex j. (If there is more than one shortest path from vertex i to vertex j, then there may be more than one distinct penultimate vertices. In this case, let p_{ij} denote the set of all penultimate vertices. If, however, we are only interested in determining one shortest path from i to j, then we need only record one vertex for p_{ij}.) If p_{ij} is known for all vertices i and j, then all the vertices along a shortest path from i to j can be found as follows: Suppose that the penultimate vertex on a shortest path from i to j is vertex k, that is, $p_{ij} = k$. Then, the second-to-last vertex on this path is the penultimate vertex on a shortest path from i to k, that is, p_{ik}. This process can be repeated until all the vertices on this path from i to j have been traced back. Hence, we need only know all p_{ij} to determine the actual shortest paths.

There are two methods to determine the p_{ij}:

1. *Tentative Method:* Tentatively set p_{ij} equal to i for all j. Do this for all vertices i. Then, as the Floyd algorithm is performed, note whenever the minimum on the right side of equation (3) is $d_{i,m}^{m-1} + d_{m,j}^{m-1}$ rather than $d_{i,j}^{m-1}$. When this occurs, set p_{ij} equal to p_{mj}. Otherwise, leave p_{ij} unchanged. [If there is a tie in equation (3), then both p_{mj} and the current value for p_{ij} may be recorded.] Upon termination of the Floyd algorithm, p_{ij} is the true penultimate vertex on the shortest path(s) from i to j.

2. *Terminal Method:* After the Floyd algorithm has terminated, vertex p_{ij} is found as follows: Vertex p_{ij} is any vertex k such that $d_{ik}^N + d_{kj}^0 = d_{ij}^N$. Only the D^0 and D^N matrices are required to determine the value(s) of k.

If the Floyd algorithm has already been performed, then one is forced to use the terminal method. However, if the Floyd algorithm is yet to be performed, then it is, of course, better to use the tentative method since this method requires little extra work when appended to the Floyd algorithm.

Another algorithm to find a shortest path between each pair of vertices in a graph was proposed by Dantzig (1967). This algorithm is similar to the Floyd shortest path algorithm in that the same calculations are performed. However, the order in which the calculations are performed is different in the *Dantzig shortest path algorithm*.

Again, number the vertices 1, 2, ..., N, and let d_{ij}^m denote the length of a shortest path from vertex i to vertex j that allows only the first m vertices to be intermediate vertices. Now, in contrast to the Floyd shortest path algorithm, let D^m be a m × m matrix whose i,j-th element is d_{ij}^m for m = 1, 2, ..., N. As before, we wish to calculate D^N the matrix whose i,j-th element denotes the length of a shortest path from vertex i to vertex j. As with the Floyd algorithm, the Dantzig algorithm calculates D^1 from D^0, etc., until D^N has been calculated from D^{N-1}.

What rationale underlies these calculations? First, note that each new matrix D^m contains one more row and one more column than its predecessor D^{m-1}. The $(m - 1)^2$ elements of D^m that are also present in D^{m-1} are calculated from D^{m-1} as done by the Floyd algorithm. The new elements d_{ij}^m, i = m or j = m, are calculated as follows: A shortest path from i to m (or from m to i) that allows only the first m vertices as intermediate vertices need never use vertex m as an intermediate vertes since all circuits have nonnegative length. Hence, a shortest path from i to m that allows only the first m vertices as intermediate vertices is any shortest path formed

3.2 All Shortest Path Algorithms

by taking a shortest path from i to some vertex j < m that uses only the first m - 1 vertices as intermediate vertices joined with the shortest arc from j to m (if such an arc exists). Similarly, a shortest path from m to i that allows only the first m vertices to be intermediate vertices is the shortest path formed by joining a shortest arc from m to some vertex j, j < m (if such an arc exists) to a shortest path from j to i that uses only the first m - 1 vertices as intermediate vertices. Lastly, let $d_{mm}^m = 0$.

With these ideas in mind, we can now formally state the *Dantzig shortest path algorithm*.

Dantzig Shortest Path Algorithm

Step 1: Number the vertices of the graph 1, 2, ..., N. Determine the matrix D^0 whose i,j-th element d_{ij}^0 equals the length of the shortest arc from vertex i to vertex j. If no such arc exists, let $d_{ij}^0 = \infty$. Let D^m be a m × m matrix whose i,j-th element is denoted by d_{ij}^m, for m = 1, 2, ..., N.

Step 2: For m = 1, 2, ..., N, successively determine each element of D^m from the elements of D^{m-1} as follows:

$$d_{mj}^m = \min_{i=1,2,\ldots,m-1} \{d_{mi}^0 + d_{ij}^{m-1}\} \quad (j = 1, 2, \ldots, m - 1) \quad (4)$$

$$d_{im}^m = \min_{j-1,2,\ldots,m-1} \{d_{ij}^{m-1} + d_{jm}^0\} \quad (i = 1, 2, \ldots, m - 1) \quad (5)$$

$$d_{ij}^m = \min\{d_{im}^m + d_{mj}^m, d_{ij}^{m-1}\} \quad (i,j = 1, 2, \ldots, m - 1) \quad (6)$$

and lastly,

$$d_{ii}^m = 0 \quad \text{(for all i and m)} \quad (7)$$

The paths corresponding to each element d_{ij}^N in D^N can be determined as before.

How many operations does the Dantzig algorithm require? To answer this question, note that the Dantzig algorithm performs essentially the same operations as the Floyd algorithm except in a different sequence. Equation (3) of the Floyd algorithm is identical to equation (6) of the Dantzig algorithm. Equations (4) and (5) of the Dantzig algorithm are simply m - 1 repetitions of equation (3) of the Floyd algorithm. Hence, both algorithms require at most the same number of calculations.

EXAMPLE 3 (Dantzig Shortest Path Algorithm). We shall perform the Dantzig algorithm to find a shortest path between each pair of vertices in the graph in Fig. 3.4. The matrix of shortest arc lengths for this graph is

$$D^0 = \begin{bmatrix} 0 & 1 & 2 & 1 \\ 2 & 0 & 7 & \infty \\ 6 & 5 & 0 & 2 \\ 1 & \infty & 4 & 0 \end{bmatrix}$$

Clearly, $D^1 = [d_{11}^1] = [0]$. The elements of D^2 are calculated as follows:

	Corresponding Path
$d_{11}^2 = 0$	
$d_{12}^2 = \min\{d_{11}^1 + d_{12}^0\} = 0 + 1 = 1$	(1,2)
$d_{22}^2 = 0$	
$d_{21}^2 = \min\{d_{21}^0 + d_{11}^1\} = 2 + 0 = 2$	(2,1)

Note that these results are identical to those in the 2 × 2 upper-left corner of D^2 calculated by the Floyd algorithm. The elements of D^3 are calculated as follows:

	Corresponding Path
$d_{13}^3 = \min\{d_{11}^2 + d_{13}^0, d_{12}^2 + d_{23}^0\} = \min\{0 + 2, 1 + 7\} = 2$	(1,3)

(continued)

3.2 All Shortest Path Algorithms

	Corresponding Path
$d_{23}^3 = \min\{d_{21}^2 + d_{13}^0,\ d_{22}^2 + d_{23}^0\} = \min\{2 + 2,\ 0 + 7\} = 4$	(2,1), (1,3)
$d_{31}^3 = \min\{d_{31}^0 + d_{11}^2,\ d_{32}^0 + d_{21}^2\} = \min\{6 + 0,\ 5 + 2\} = 6$	(3,1)
$d_{32}^3 = \min\{d_{32}^0 + d_{22}^2,\ d_{31}^0 + d_{12}^2\} = \min\{5 + 0,\ 6 + 1\} = 5$	(3,2)
$d_{12}^3 = \min\{d_{12}^2,\ d_{13}^3 + d_{32}^3\} = \min\{1,\ 2 + 5\} = 1$	(1,2)
$d_{21}^3 = \min\{d_{21}^2,\ d_{23}^3 + d_{31}^3\} = \min\{2,\ 4 + 6\} = 2$	(2,1)
$d_{11}^3 = d_{22}^3 = d_{33}^3 = 0$	

Thus,

$$D^3 = \begin{bmatrix} 0 & 1 & 2 \\ 2 & 0 & 4 \\ 6 & 5 & 0 \end{bmatrix}$$

and the shortest paths corresponding to D^3 are

$$\begin{bmatrix} & (1,2) & (1,3) \\ (2,1) & & (2,1),\ (1,3) \\ (3,1) & (3,2) & \end{bmatrix}$$

Note that the matrix D^3 computed here is identical to the 3 × 3 upper-left corner of the matrix D^3 computed by the Floyd shortest path algorithm. It is left to the interested reader to repeat these calculations for D^4 and to verify that the matrix D^4 computed by the Dantzig algorithm is identical to the matrix D^4 computed by the Floyd algorithm.

Since the terminal method for constructing the arcs in a shortest path uses only the D^0 and D^N matrices, the terminal method is the same for both the Floyd and Dantzig algorithms.

However, the tentative method for constructing the arcs in a shortest path for the Dantzig algorithm is slightly different than the tentative method for the Floyd algorithm. Specifically, whenever equation (4) is performed in the Dantzig algorithm, let p_{mj} be the vertex that determines the right-side minimization. Whenever equation (5) is performed in the Dantzig algorithm, let p_{im} be the vertex that determines the right-side minimization. Equation (6) of the Dantzig algorithm is identical to equation (3) of the Floyd algorithm, and the tentative method remains the same, namely, let $p_{ij} = m$ if the minimum of the right side is $d_{im}^m + d_{mj}^m$. Otherwise, the minimum of the right side is d_{ij}^{m-1} and p_{ij} must remain unchanged.

If we want to find a shortest path between every pair of vertices in a graph, happily, we have a choice of several algorithms. The Floyd (or, equivalently Dantzig) algorithm could be used, or the Dijkstra algorithm could be repeated once for each vertex as the initial vertex. Consequently, it becomes necessary to compare the number of arithmetic operations required by the various algorithms. The study of the number of operations required by an algorithm is known as *computational complexity*.

For a given graph, the computational complexity of some algorithms is easy to evaluate because the number of operations required by the algorithm is not subject to any variation. The Dijkstra, Floyd, and Dantzig algorithms fall into this category. For other algorithms, the exact number of operations cannot be determined in advance. For example, the exact number of operations required by the Ford algorithm cannot be determined in advance since there is no way to know how many times each vertex will be colored. For algorithms in this category, it is customary to calculate the worst possible number of operations required for completion.

As we have seen, all shortest path algorithms consist of essentially two arithmetic operations, addition and minimization. To analyze the computational complexity of one of these algorithms, we need some way of comparing addition operations with minimization operations. Of course, this comparison varies between computers (human and mechanical), but for expediency we shall assume that these two operations require equivalent amounts of computational time.

3.2 All Shortest Path Algorithms 63

The Floyd algorithm must compute N matrices $D^1, D^2, \ldots, D^N$. Each of these matrices consists of N^2 elements. Hence, a total of N^3 elements must be computed by the Floyd algorithm. Each of these computations requires by equation (3) one addition and one minimization. Hence, the Floyd algorithm requires roughly N^3 additions and N^3 minimizations. [Strictly speaking, this is an overstatement since some elements of D^i can be taken directly from D^{i-1} without performing equation (3) as the elements i-th row and i-th column of D^{i-1} and D^i are identical.] The total amount of computation required by the Floyd algorithm is proportional to $2N^3$. Or, in more technical terminology, the Floyd algorithm requires $O(2N^3)$ running time.

Next, let us consider the computational complexity of the Dijkstra algorithm. At the first iteration of the Dijkstra algorithm, the N - 1 uncolored vertices must be examined. From equation (1), this requires N - 1 additions, N - 1 minimizations, and the selection of the smallest of N - 1 numbers, i.e., another N - 1 minimizations. Thus, 3(N - 1) operations are required by the first iteration. Similarly, 3(N - 2) operations are required by the second iteration, etc. In total, $\sum_{i=1}^{i=N} 3(N - i) = 3N(N - 1)/2$ operations are required. Of course, at each iteration one must also determine which vertices are colored and which are uncolored. This requires additional work, but a clever programming technique can be used to avoid this. The details of this are beyond the scope of this text but can be found in Yen (1973) and Williams and White (1973). Thus, the Dijkstra algorithm requires $O(1\ 1/2\ N^2)$ running time. From this, it follows that the Ford algorithm requires at worst $O(1\ 1/2\ N^3)$ running time since each vertex may be colored as many as N - 1 times, whereas the Dijkstra algorithm colors each vertex at most once.

To summarize,

Dijkstra algorithm requires $1\ 1/2\ N^2$ operations.
Ford algorithm requires at worst $1\ 1/2\ N^3$ operations.
Floyd algorithm requires $2N^3$ operations.
Dantzig algorithm requires $2N^3$ operations.

As suggested, the Floyd algorithm could be replaced by repeating the Dijkstra algorithm N times, once for each vertex as the initial vertex. This requires $O(1\ 1/2\ N^3)$ running time which is superior to the $O(2N^3)$ running time of the Floyd algorithm. If, however, some arc lengths are negative (but, of course, there are no negative circuits), then the Ford algorithm must replace the Dijkstra algorithm and at worst the running time is $O(1\ 1/2\ N^4)$ which is inferior for large N to the Floyd algorithm running time of $O(2N^3)$. However, it must not be forgotten that most likely the Ford algorithm will terminate with far less than the worst possible number of operations.

Computational complexity is a growing area attracting much research interest. The interested reader is directed to the survey articles of Karp (1975) and Lawler (1971), and to Knuth (1973), a valuable reference for those interested in efficient programming techniques applicable to the algorithm described here.

3.3 THE K-TH SHORTEST PATH ALGORITHM

The two preceding sections considered the problem of finding various shortest paths. Often, however, knowledge of the second, third, fourth, etc., shortest path between two vertices is useful. For example, the airline in Example 2 of Sec. 3.1 might want to know the runner-up shortest flight routes between Springfield and Ankara just in case one of its clients cannot take the shortest flight route due to visa difficulties, flight cancellations, or airlines strikes along the shortest flight route.

This section first presents an algorithm, called the *Double sweep algorithm*, that finds the k shortest path lengths between a specified vertex and all other vertices in the graph. Next, this section presents two algorithms, called the *Generalized Floyd algorithm* and the *Generalized Dantzig algorithm*, that find the k shortest path lengths between every pair of vertices in the graph.

The Dijkstra, Floyd, and Dantzig algorithms of Sec. 3.1 and 3.2 were able to construct various shortest paths. These algorithms essentially consisted of performing a sequence of two arithmetic

3.3 The K-th Shortest Path Algorithm

operations, addition and minimization. These two operations were performed on single numbers that represented either arc lengths or path lengths. For example, equation (1), which defines the Dijkstra algorithm, consists exclusively of addition and minimization. The same is true for equation (3), which defines the Floyd algorithm, and equations (4)-(7), which define the Dantzig algorithm.

The algorithms to be presented in this section (double-sweep algorithm, generalized Floyd algorithm, and generalized Dantzig algorithm) also consist exclusively of addition and minimization operations. However, these operations are performed not on single numbers (as with the previous algorithms) but on sets of k distinct numbers that represent the lengths of paths or arcs. With this as motivation, let R^k denote the set of all vectors $(d_1, d_2, \ldots, d_k)$ with the property that $d_1 < d_2 < \ldots < d_k$.† Thus, the components of a member of R^k are distinct and arranged in ascending order. For example, $(-3, -1, 0, 4, 27) \in R^5$.

Let $A = (a_1, a_2, \ldots, a_k)$ and $B = (b_1, b_2, \ldots, b_k)$ be two members of R^k. *Generalized minimization*, denoted by +, is defined as

$$A + B = \min_k \{a_i, b_i: i = 1, 2, \ldots, k\} \tag{8}$$

where $\min_k(X)$ means the k smallest distinct members of the set X.

Generalized addition, denoted by ×, is defined as

$$A \times B = \min_k \{a_i + b_j: i,j = 1, 2, \ldots, k\} \tag{9}$$

For example, if $A = (1, 3, 4, 8)$ and $B = (3, 5, 7, 16)$, then $A + B = \min_4(1, 3, 4, 8, 3, 5, 7, 16) = (1, 3, 4, 5)$ and $A \times B = \min_4(1 + 3, 1 + 5, 1 + 7, 1 + 16, 3 + 3, 3 + 5, 3 + 7, 3 + 16, 4 + 3, \ldots) = (4, 6, 7, 8)$. Note that the components of $A + B$ and $A \times B$ can be arranged so that $A + B \in R^k$ and $A \times B \in R^k$. Moreover, since members of R^k have their components arranged in ascending order, generalized minimization need not require more than k comparisons and generalized addition need not require more than k(k - 1) additions and k(k - 1) comparisons.

---
†Let $\infty \leq \infty$.

Extending our previous notation, let $d_{ij}^0 = (d_{ij1}^0, d_{ij2}^0, \ldots, d_{ijk}^0) \in R^k$ denote the lengths of the k shortest arcs from vertex i to vertex j. If two arcs from vertex i to vertex j have the same length, then this length appears only once in d_{ij}^0. If there are less than k arcs from i to j, then fill up the remaining components with ∞. For example, if vertex 5 is joined to vertex 3 by three arcs of lengths 9, 13 and 9, then for k = 4, $d_{53}^0 = (9, 13, \infty, \infty)$. If i = j, then suppose that there is an arc from vertex i to itself with length zero. Let D^0 denote the matrix whose i,j-th element is d_{ij}^0.

Let $d_{ij}^m = (d_{ij1}^m, d_{ij2}^m, \ldots, d_{ijk}^m) \in R^k$ denote the k shortest distinct path lengths from vertex i to vertex j that use only vertices 1, 2, ..., m as intermediate vertices. (Recall that in Sec. 3.1 the vertices were numbered 1, 2, ..., N.) Let D^m denote the matrix whose i,j-th element is d_{ij}^m.

Lastly, let $d_{ij}^* = (d_{ij1}^*, d_{ij2}^*, \ldots, d_{ijk}^*) \in R^k$ denote the k shortest distinct path lengths from vertex i to vertex j. These path lengths are called the *optimal path lengths*. Let D^* denote the matrix whose i,j-th element is d_{ij}^*.

Let matrix L be formed from matrix D^0 by replacing every component of every element d_{ij}^0 by ∞ whenever $i \leq j$. Let matrix U be formed from matrix D^0 by replacing every component of every element d_{ij}^0 by ∞ whenever $i \geq j$. Matrices L and U are called the upper and lower triangular portions of D^0. For example, for k = 2 and

$$D^0 = \begin{bmatrix} (0, 1) & (6, 10) & (2, 9) \\ (1, 8) & (0, \infty) & (3, \infty) \\ (\infty, \infty) & (2, 4) & (0, \infty) \end{bmatrix}$$

$$L = \begin{bmatrix} (\infty, \infty) & (\infty, \infty) & (\infty, \infty) \\ (1, 8) & (\infty, \infty) & (\infty, \infty) \\ (\infty, \infty) & (2, 4) & (\infty, \infty) \end{bmatrix}$$

and

$$U = \begin{bmatrix} (\infty, \infty) & (6, 10) & (2, 9) \\ (\infty, \infty) & (\infty, \infty) & (3, \infty) \\ (\infty, \infty) & (\infty, \infty) & (\infty, \infty) \end{bmatrix}$$

3.3 The K-th Shortest Path Algorithm

If we wish to know only the k shortest path lengths from a specific vertex (say vertex 1) to every other vertex in the graph, then we need only determine the first row $(d_{11}^*, d_{12}^*, \ldots, d_{1N}^*)$ of D^*. Note that this row consists of N members each belonging to R^k. Hence, altogether Nk values and the corresponding paths must be determined.

The double-sweep algorithm (Shier, 1974; 1976) is an efficient way of computing these Nk paths. The double-sweep algorithm will work as long as the graph contains no circuits with negative length. The double-sweep algorithm is initiated with any estimate $(d_{11}^{(0)}, d_{12}^{(0)}, \ldots, d_{1N}^{(0)})$ of $(d_{11}^*, d_{12}^*, \ldots, d_{1N}^*)$ in which each value is not less than the corresponding optimal value and for which $d_{111}^{(0)} = 0$. For example, all values could be set equal to ∞ except $d_{111}^{(0)}$ which equals zero. The algorithm then computes new improved (i.e., reduced) estimates of d_{1i}^* by seeing if any of the k values in $d_{1j}^{(0)} \times d_{ji}^0$ are less than any of the k values in the current estimate $d_{1i}^{(0)}$ of d_{1i}^*. If so, the smallest k values are chosen. This process is repeated for all j yielding a new estimate $d_{1i}^{(1)}$ of d_{1i}^*. The algorithm terminates when two successive estimates $(d_{11}^{(i)}, d_{12}^{(i)}, \ldots, d_{1N}^{(i)})$ and $(d_{11}^{(i+1)}, d_{12}^{(i+1)}, \ldots, d_{1N}^{(i+1)})$ are identical in every component for $i \geq 1$. The terminal values of the estimates can be shown to equal the optimal values. See the proof of the double-sweep algorithm.

The double-sweep algorithm has the added efficiency that whenever possible in the revision of an estimate vector $d_1^{(i)} = (d_{11}^{(i)}, d_{12}^{(i)}, \ldots, d_{1N}^{(i)})$ to the next estimate $d_1^{(i+1)} = (d_{11}^{(i+1)}, d_{12}^{(i+1)}, \ldots, d_{1N}^{(i+1)})$, the values already computed in $d_1^{(i+1)}$ are used in the computation of the remaining values in $d_1^{(i+1)}$. This immediate use of updated estimate values can accelerate the discovery of the optimal values. With these ideas in mind, we can now formally state the double-sweep algorithm.

Double Sweep Algorithm

Step 1: Initialization. Let the initial estimate $d_1^{(0)} = d_{11}^{(0)}, d_{12}^{(0)}, \ldots, d_{1N}^{(0)}$) of d_i^* consist of values that equal or exceed the corresponding optimal values. However, let $d_{111}^{(0)} = 0$ since there is a path of zero length (the path without any arcs) from vertex 1 to itself.

Step 2: Given an estimate $d_1^{(2r)}$ of d_1^*, calculate new estimates $d_1^{(2r+1)}$ and $d_1^{(2r+2)}$ as follows:

Backward sweep:

$$d_1^{(2r+1)} = d_1^{(2r+1)} L + d_1^{(2r)} \qquad (10)$$

Forward sweep:

$$d_1^{(2r+2)} = d_1^{(2r+2)} U + d_1^{(2r+1)} \qquad (r = 0, 1, 2, \ldots) \qquad (11)$$

Note that the addition and multiplication operations in the above matrix multiplications refer to generalized minimization and generalized addition, respectively, as defined by equations (8) and (9).

Terminate when two successive estimates $d_1^{(t-1)}$ and $d_1^{(t)}$ are identical for $t > 1$. The terminal estimate $d_1^{(t)}$ equals d_i^*. Stop.

At this point, the reader might wonder how to perform the computations required by equations (10) and (11) since the vector to be computed appears on both sides of each equation. For equation (10), first determine the last component $d_{1N}^{(2r+1)}$ of $d_1^{(2r+1)}$. (This is possible since the N-th column of L consists entirely of infinite entries, and hence $d_1^{(2r+1)}$ is not needed to compute $d_{1N}^{(2r+1)}$.) Next, determine the second to last component $d_{1,N-1}^{(2r+1)}$ of $d_1^{(2r+1)}$. [This is possible since the (N-1)th column of L contains infinite entries except in its last row, and consequently only $d_{1N}^{(2r+1)}$, which has already been calculated, is needed to determine $d_{1,N-1}^{(2r+1)}$.] Next, determine $d_{1,N-2}^{(2r+1)}$, etc. Hence, equation (10) computes the components of $d_1^{(2r+1)}$ from last to first and is called the *backward sweep*. In a similar way, equation (11) computer the components of $d_1^{(2r+2)}$ from first to last and is called the *forward sweep*.

3.3 The K-th Shortest Path Algorithm

The double-sweep algorithm determines path lengths. How can the actual path (or paths) corresponding to the path length be found? Since the double-sweep algorithm is initialized with arbitrary larger-than-optimal estimates that do not correspond to any particular path, we cannot carry along at each iteration the path (or paths) corresponding to each new estimate as was possible with the Floyd algorithm and Dantzig algorithm. The path corresponding to each path length can be retrieved as follows: Suppose we wish to find the m-th shortest path from vertex 1 to vertex j. This path must consist of the ℓ-th ($\ell \leq m$) shortest path from vertex 1 to some vertex i together with an arc from vertex i to vertex j. The length of this path plus the length of this arc must equal d_{1jm}^{*}. Vertex i is called the penultimate vertex on the m-th shortest path from vertex 1 to vertex j. Vertex i is determined searching through the arc length matrix D^0 and the set of shortest path lengths d_1^{*}. Once vertex i has been located, this process can be repeated to find the penultimate vertex on the ℓ-th shortest path from vertex 1 to vertex i. Ultimately, the entire m-th shortest path from vertex 1 to vertex j can be traced back by repeating this process.

It is possible that the penultimate vertex is not unique. In this case, all penultimate vertices can be recorded and all the paths of equal length that tie for the m-th shortest path from vertex 1 to vertex j can be traced back (if desired).

If all paths are to be determined, it is computationally best to determine first all shortest paths, then all second shortest paths, etc., since knowledge of the ℓ-th shortest paths is needed to determine the m-th ($\ell \leq m$) shortest paths.

Proof of the double-sweep algorithm: Equation (10) is a generalized minimization between corresponding components of $d_1^{(2r+1)}$ L and $d_1^{(2r)}$. Similarly, equation (11) is a generalized minimization between corresponding components of $d_1^{(2r+2)}$ U and $d_1^{(2r+1)}$. Hence, no value is ever replaced with a larger value in the succeeding estimate. Thus, $d_1^{(0)}$, $d_1^{(1)}$, ..., $d_1^{(t)}$ from Nk nonincreasing sequences. Moreover, no value can ever be less than its optimal value (which we will assume to be finite or +∞) for the following reason: Suppose that the *first* less-than-optimal value to be calculated is for the m-th shortest path from vertex 1 to vertex j. By equations (10) and (11), this less-than-optimal path length was computed as the sum of a number not less than the length of a shortest path from vertex 1 to some vertex i, plus the length of an arc from vertex i to vertex j. This implies that this path length is less than optimal which contradicts our assumption that we are computing the first less-than-optimal path length to appear.

Thus, the sequence of estimates $d_1^{(0)}$, $d_1^{(1)}$, ..., $d_1^{(t)}$ forms Nk nonincreasing sequences of path lengths that are always greater than or equal to their corresponding optimal values. It remains to show that each of these Nk sequences converges to its optimal value in a finite number of steps. This will be accomplished by showing that each succeeding estimation in this sequence contains at least one more optimal value than its predecessor.

First, some lemmas are needed.

LEMMA 3.1. If vertex i is the penultimate vertex of some m-th shortest path from vertex 1 to vertex j, then the portion P' of P from vertex 1 to vertex i is one of the m shortest paths from vertex 1 to vertex i.

Proof: If path P' were not among the m shortest paths from vertex 1 to vertex i, then there would be m paths from vertex 1 to vertex i with distinct lengths that are shorter than P'. Each of these paths could be extended to vertex j by the addition of a single arc. This would result in m paths from vertex 1 to vertex j that are shorter than path P, which is a contradiction. Q.E.D.

LEMMA 3.2. If the m shortest path lengths from vertex 1 to every other vertex are known, then the length of the (m + 1)-st shortest path length from vertex 1 to itself is determined during the next double sweep for $m \geq 0$.

Proof: Consider any (m + 1)-st shortest path P from vertex 1 to itself. Let i denote the penultimate vertex of path P. From

3.3 The K-th Shortest Path Algorithm

Lemma 1, it follows that the portion P' of P from vertex 1 to vertex i is among the (m + 1)-st shortest paths from vertex 1 to vertex i. Path P' cannot be an (m + 1)-st shortest path otherwise, there would be m + 1 shortest paths from vertex 1 to itself that have vertex i as their penultimate vertex which contradicts the fact that $d^*_{111} = 0$ and corresponds to a path without arcs.

By assumption, the m shortest path lengths from vertex 1 to vertex i are known. Also, the lengths of the k shortest arcs from vertex i to vertex 1 are known. Hence, the length of P will be determined during the next double sweep. Q.E.D.

We shall now employ these two lemmas to show that at each double-sweep at least one more optimal value is calculated. Suppose that prior to the r-th double sweep, the length of the m shortest paths from vertex 1 to every other vertex has been determined for some $m \geq 0$. Let j be any vertex for which the optimal value $d^*_{1j,m+1}$ of the (m + 1)-st shortest path from vertex 1 to vertex j has not yet been determined.

Let i denote the penultimate vertex of some (m + 1)-st shortest path from vertex 1 to vertex j. If the length of the (m + 1)-st shortest path from vertex 1 to vertex i has already been determined, then the next double sweep will determine the optimum value of the length of the (m + 1)-st shortest path from vertex 1 to vertex j since this path is the merger of one of the (m + 1)-st shortest paths from vertex 1 to vertex i and an arc from i to j.

If the length of the (m + 1)-st shortest path from vertex 1 to vertex i has not yet been determined by the algorithm, then we cannot be certain that the next double sweep will generate the optimum value of the length of the (m + 1)-st shortest path from vertex 1 to vertex j. If the next double sweep does in fact generate this path length, then the proof is complete. If not, then the (m + 1)-st shortest path from vertex 1 to vertex j must consist of the (m + 1)-st shortest path from 1 to i together with an arc from i to j, by Lemma 3.1.

Repeat the preceding argument replacing vertex i by vertex j. Successive repetitions of this argument will either lead to a situation in which an optimum value is calculated during the next double-sweep iteration of the algorithm or else the (m + 1)-st shortest path from vertex 1 to vertex j will be traced back to vertex 1. If the latter occurs, then by Lemma 3.2, the next double sweep determines an optimal value for the (m + 1)-st shortest path from vertex 1 to itself, which concludes the proof that the algorithm terminates finitely. Q.E.D.

As a postscript to the proof, note that if the graph contained no circuits with negative length that are accessible from vertex 1, then after $r \geq N$ double sweeps, the first component of each estimate vector $d_{11}^{(2r)}$, $d_{12}^{(2r)}$, ..., $d_{1N}^{(2r)}$ would no longer decrease. Similarly, after $r \geq 2N$ double seeeps the first two components of each estimate vector $d_{11}^{(2r)}$, $d_{12}^{(2r)}$, ..., $d_{1N}^{(2r)}$ would no longer decrease. And in general, after $r \geq cN$ double sweeps, the first c components of each estimate vector $d_{11}^{(2r)}$, $d_{12}^{(2r)}$, ..., $d_{1N}^{(2r)}$ would no longer decrease. (Note that the superscript 2r is used since each double sweep determines two estimates.)

Hence, we can detect the existence of a circuit with negative length that is accessible from vertex 1 by noticing any decrease (or for that matter any change at all) in the first c components of any estimate vector $d_{11}^{(2r)}$, $d_{12}^{(2r)}$, ..., $d_{1N}^{(2r)}$ after the r-th double sweep ($r \geq c$) for any c = 1, 2, ..., N.

Since the double-sweep algorithm calculates at least one additional optimal value at each double sweep, at most Nk single sweeps are required. In practice, far fewer sweeps are required since more than one optimal value is frequently computed during one double sweep.

How many generalized additions and generalized minimizations does a double sweep require? There are $N^2 - N$ elements off the diagonal of matrix D^0. Hence, matrices L and U each have $1/2(N^2 - N)$ elements that are possibly finite. Each of these elements requires one generalized addition and one generalized minimization. Thus, each double sweep requires roughly N^2 generalized additions and N^2 generalized minimizations. Hence, at most $1/2kN^3$ generalized additions and minimizations are required to complete the double-sweep algorithm.

For k = 1, the double-sweep algorithm requires at most $1/2N^3$ additions and $1/2N^3$ minimizations. Thus, for k = 1, the double-sweep

3.3 The K-th Shortest Path Algorithm

algorithm requires $O(N^3)$ running time which is superior to the worst running time of $O(1\ 1/2N^3)$ for the Ford algorithm but inferior to the running time of $O(1\ 1/2N^2)$ for the Dijkstra algorithm. Of course, all comparisons must be tempered with the reminder that we are considering the worst possible running times for the double-sweep and Ford algorithms.

EXAMPLE 1 (Double-Sweep Algorithm). Let us use the double-sweep algorithm to find the lengths of the three shortest paths from vertex 1 to every other vertex in the graph in Fig. 3.5. The matrix D^0 of path lengths is

$$\begin{bmatrix} (0, \infty, \infty) & (1, \infty, \infty) & (\infty, \infty, \infty) & (\infty, \infty, \infty) \\ (2, 4, \infty) & (0, \infty, \infty) & (-1, \infty, \infty) & (\infty, \infty, \infty) \\ (2, \infty, \infty) & (\infty, \infty, \infty) & (0, \infty, \infty) & (-1, \infty, \infty) \\ (\infty, \infty, \infty) & (3, \infty, \infty) & (2, \infty, \infty) & (0, \infty, \infty) \end{bmatrix}$$

Matrix L is

$$\begin{bmatrix} (\infty, \infty, \infty) & (\infty, \infty, \infty) & (\infty, \infty, \infty) & (\infty, \infty, \infty) \\ (2, 4, \infty) & (\infty, \infty, \infty) & (\infty, \infty, \infty) & (\infty, \infty, \infty) \\ (2, \infty, \infty) & (\infty, \infty, \infty) & (\infty, \infty, \infty) & (\infty, \infty, \infty) \\ (\infty, \infty, \infty) & (3, \infty, \infty) & (2, \infty, \infty) & (\infty, \infty, \infty) \end{bmatrix}$$

Matrix U is

$$\begin{bmatrix} (\infty, \infty, \infty) & (1, \infty, \infty) & (\infty, \infty, \infty) & (\infty, \infty, \infty) \\ (\infty, \infty, \infty) & (\infty, \infty, \infty) & (-1, \infty, \infty) & (\infty, \infty, \infty) \\ (\infty, \infty, \infty) & (\infty, \infty, \infty) & (\infty, \infty, \infty) & (-1, \infty, \infty) \\ (\infty, \infty, \infty) & (\infty, \infty, \infty) & (\infty, \infty, \infty) & (\infty, \infty, \infty) \end{bmatrix}$$

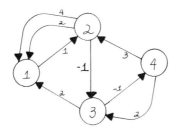

Figure 3.5
Double-Sweep Algorithm

From equation (10), the backward sweep becomes:

$$d_{14}^{(2r+1)} = d_{14}^{(2r)}$$

$$d_{13}^{(2r+1)} = (2, \infty, \infty) \times d_{14}^{(2r+1)} + d_{13}^{(2r)}$$

$$d_{12}^{(2r+1)} = (3, \infty, \infty) \times d_{14}^{(2r+1)} + d_{12}^{(2r)}$$

$$d_{11}^{(2r+1)} = (2, 4, \infty) \times d_{12}^{(2r+1)} + (2, \infty, \infty) \times d_{13}^{(2r+1)} + d_{11}^{(2r)}$$

From equation (11), the forward sweep becomes

$$d_{11}^{(2r+2)} = d_{11}^{(2r+1)}$$

$$d_{12}^{(2r+2)} = (1, \infty, \infty) \times d_{11}^{(2r+2)} + d_{12}^{(2r+1)}$$

$$d_{13}^{(2r+2)} = (-1, \infty, \infty) \times d_{12}^{(2r+2)} + d_{13}^{(2r+1)}$$

$$d_{14}^{(2r+2)} = (-1, \infty, \infty) \times d_{13}^{(2r+2)} + d_{14}^{(2r+1)}$$

The following table gives the results of the first five sweeps. The results of the fourth and fifth sweeps are identical in every component and hence represent the optimal path lengths.

	$d_{11}^{(r)}$	$d_{12}^{(r)}$	$d_{13}^{(r)}$	$d_{14}^{(r)}$	
r = 0	(0, ∞, ∞)	(∞, ∞, ∞)	(∞, ∞, ∞)	(∞, ∞, ∞)	
					← (backward)
r = 1	(0, ∞, ∞)	(∞, ∞, ∞)	(∞, ∞, ∞)	(∞, ∞, ∞)	
					← (forward)
r = 2	(0, ∞, ∞)	(1, ∞, ∞)	(0, ∞, ∞)	(-1, ∞, ∞)	
					← (backward)
r = 3	(0, 2, 3)	(1, 2, ∞)	(0, 1, ∞)	(-1, ∞, ∞)	
					← (forward)
r = 4	(0, 2, 3)	(1, 2, 3)	(0, 1, 2)	(-1, 0, 1)	
					← (backward)
r = 5	(0, 2, 3)	(1, 2, 3)	(0, 1, 2)	(-1, 0, 1)	

Note that the double-sweep algorithm terminated after 5 sweeps or 2 1/2 double sweeps compared to the theoretical maximum of 1/2Nk = 6 double sweeps.

In the remainder of this section, we shall consider the more general problem of finding the k shortest paths from *each* vertex to

3.3 The K-th Shortest Path Algorithm

every other vertex. The Floyd algorithm and Dantzig algorithm of Sec. 3.2 solved this problem for k = 1. These algorithms will now be extended to k > 1.

Of course, this problem could be solved by performing the double-sweep algorithm once for each vertex as the initial vertex. However, computationally this is not very efficient since much information regarding shortest paths from a vertex that is not the initial vertex would never be utilized.

Before proceeding to the generalized Floyd algorithm and the generalized Dantzig algorithm for finding the k shortest paths between every pair of vertices, an additional definition is needed.

Let $d_{ii} \in R^k$ denote the lengths of any k paths from vertex i to itself including the empty path of length zero. These paths are circuits. The *convolution* d_{ii}^C of d_{ii} is defined as the lengths of the k distinct shortest paths from vertex i to itself that can be formed by combining paths represented by a component of d_{ii}. Thus, each member of d_{ii}^C corresponds to a path formed by repeating one or more of the circuits in d_{ii}. Mathematically,

$$d_{ii}^C = d_{ii} \times d_{ii} \times d_{ii} \times \ldots \times d_{ii} \in R^k \qquad (12)$$

(k times)

(Recall that × denotes generalized addition.)

The ideas underlying the generalized Floyd algorithm and the generalized Dantzig algorithm are identical to those underlying the Floyd algorithm and the Dantzig algorithm, except that addition is replaced by generalized addition and minimization is replaced by generalized minimization. Also, one must consider the possibility of generating paths that begin with a circuit attached to the initial vertex or terminate with a circuit attached to the terminal vertex.

The generalized Floyd algorithm is the same as the original Floyd algorithm except that d_{ij}^m now represents a vector in R^k signifying the lengths of the k shortest paths from vertex i to vertex j that use only the first m vertices as intermediate vertices. Equation (3) becomes

$$d_{mm}^m = (d_{mm}^{m-1})c \tag{13}$$

$$d_{im}^m = d_{im}^{m-1} d_{mm}^m \quad \text{(for all } i \neq m\text{)} \tag{14}$$

$$d_{mi}^m = d_{mm}^m d_{mi}^{m-1} \quad \text{(for all } i \neq m\text{)} \tag{15}$$

$$d_{ij}^m = d_{im}^m d_{mj}^m + d_{ij}^{m-1} \quad \text{(for all } i, j \neq m\text{)} \tag{16}$$

The generalized Dantzig algorithm is the same as the original Dantzig algorithm except that d_{ij}^m now represents a vector in R^k signifying the lengths of the k shortest paths from vertex i to vertex j that use only the first m vertices as intermediate vertices. Equation (7) becomes

$$d_{mm}^m = (d_{mm}^0 + \sum_{i=1}^{m-1} \sum_{j=1}^{m-1} d_{mi}^0 d_{ij}^{m-1} d_{jm}^0)c \tag{17}$$

Equation (4) becomes

$$d_{mi}^m = \sum_{j=1}^{j=m-1} d_{mm}^m d_{mj}^0 d_{ji}^{m-1} \quad (i = 1, 2, \ldots, m-1) \tag{18}$$

Equation (5) becomes

$$d_{im}^m = \sum_{j=1}^{j=m-1} d_{ij}^{m-1} d_{jm}^0 d_{mm}^m \quad (i = 1, 2, \ldots, m-1) \tag{19}$$

Equation (6) becomes

$$d_{ij}^m = d_{im}^m d_{mj}^m + d_{ij}^{m-1} \quad (i, j = 1, 2, \ldots, m-1) \tag{20}$$

Note that in each generalized algorithm d_{mm}^m is the first element of D^m to be calculated. Next, d_{mm}^m is used to calculate d_{im}^m and d_{mi}^m for all $i \neq m$. Lastly, the elements d_{ij}^m ($i \neq m$, $j \neq m$) are calculated

In the generalized Floyd algorithm, d_{mm}^m is calculated in equation (13) by taking the smallest k combinations of all simple circuits from vertex m to itself that use only the first m vertices as intermediate vertices.

3.3 The K-th Shortest Path Algorithm

Then, d_{im}^m is calculated in equation (14) by taking the smallest k combinations of paths d_{im}^{m-1} from i to m coupled with circuits d_{mm}^m at m. In a similar way, equation (15) calculates d_{mi}^m. Lastly, the calculation of d_{ij}^m in equation (16) is merely the generalized-operator restatement of equation (3).

In the generalized Dantzig algorithm, d_{mm}^m is calculated in equation (17) by taking the smallest k combinations of (a) loops from vertex m to itself, and (b) circuits that start at vertex m proceed to some other vertex i (i < m), then proceed to some other vertex j (j < m, possibly j = 1), and then return to vertex m, that is $d_{mi}^0 d_{ij}^{m-1} d_{jm}^0$ for all i, j < m. Equation (18) is equation (4) restated in terms of generalized operations and allowing for circuits d_{mm}^m attached to vertex m. Similarly, equation (19) is equation (5) restated in terms of generalized operations and allowing circuits d_{mm}^m attached to vertex m. Lastly, equation (20) is equation (6) restated in terms of generalized operations.

As with the original algorithms, the optimality of the generalized algorithms can be proved by an induction on N, the number of vertices in the graph.

Let us determine how many generalized operations are required by the generalized Floyd algorithm. The algorithm must calculate N matrices D^0, D^1, ..., D^N. Each matrix requires one performance of equation (13), N - 1 performances of equation (14), N - 1 performances of equation (15) and $(N - 1)^2$ performances of equation (16).

Equation (13) requires one convolution, which by equation (12) consists of k generalized additions. Hence, equation (13) requires a total of k generalized additions.

Equation (14) and (15) each require one generalized addition. Equation (16) requires one generalized addition and one generalized minimization.

Thus, each matrix D^1, D^2, ..., D^N requires $N^2 + k - 2$ generalized additions and $(N - 1)^2$ generalized minimizations, or roughly N^2 generalized additions and N^2 generalized minimizations. In total, the generalized Floyd algorithm requires about N^3 generalized additions and N^3 generalized minimizations. Since the generalized Dantzig algorithm is essentially a resequencing of the operations performed by the generalized Floyd algorithm, it requires approximately the same number of operations. Hence, each has a running time of $O(2N^3)$.

How does this compare with the number of operations required by the double-sweep algorithm to do the same job? As shown before, the double-sweep algorithm requires at most a running time of $O(kN^3)$, Nk generalized additions and N^3k generalized minimizations. If the double-sweep algorithm were performed N times (once for each vertex as the initial vertex), then at most a running time of $O(kN^4)$ is required. However, since the double-sweep algorithm usually terminates early, the actual running time will be much less.

Hence, we can conclude that the choice of algorithm depends upon (1) the comparison of kN^4 and $2N^3$, (2) our estimates on how prematurely the double-sweep algorithm will terminate with an optimal solution, and (3) the observation that fewer operations are required to perform a generalized operation in the double-sweep algorithm since the vectors in the double-sweep algorithm usually consists of many components euqal to infinity. Computational experience with the double-sweep algorithm has been given by Shier (1974).

3.4 OTHER SHORTEST PATHS

All the shortest path algorithms studied thus far can be regarded as well-defined sequences of addition operations and minimization operations.

Let us consider a different path problem, namely the bottleneck problem. The *bottleneck* of a path is defined as the length of the shortest arc in the path. *The bottleneck problem* is the problem of finding a path between two vertices with the largest possible bottleneck.

EXAMPLE 1. Suppose that a bridge collapses when its weight limit is exceeded. We might wish to know the maximum weight that a vehicle can carry between points a and b so that it does not exceed the weight limit of any bridge that it crosses. If we consider the bridges as

3.4 Other Shortest Paths

arcs and let the arc lengths equal the corresponding weight limits, this problem becomes the problem of finding a path from a to b with the largest bottleneck.

In the shortest path problem, we associate a number to each path called its length. This number is found by adding together all arc lengths in this path. In the bottleneck problem, we associate a number to each path called its bottleneck. This number is found by taking the minimum of all arc lengths in this path. In the shortest path problem, one path is preferred to another path if its length is smaller than the length of the other path. This requires a minimization operation. In the bottleneck problem, one path is preferred to another path if its bottleneck is larger than the bottleneck of the other path. This requires a maximization operation. Thus, the bottleneck problem can be regarded as similar to the shortest path problem except that minimization replaces addition and maximization replaces minimization.

If the Floyd algorithm or Dantzig algorithm or their generalizations, or the double-sweep algorithm were performed with traditional minimization as the addition operation and with traditional maximization as the minimization operation, then these algorithms would produce the best bottleneck paths. Moreover, if traditional minimization was retained as the minimization operation, then these algorithms would produce the worst bottleneck paths.

However, if any of these algorithms are used to solve the bottleneck problem, we must remember that the absence of an arc must be interpreted as a bottleneck of $-\infty$, whereas in the shortest path problem the absence of an arc was interpreted as a length of $+\infty$. Similarly, the null path (path without any arcs) has a bottleneck of $-\infty$, whereas the null path has a length of zero. Thus, for the bottleneck problem, the matrix D^0 that initialized the Floyd algorithm and the Dantzig algorithm consists of arc lengths and $-\infty$ wherever no arc appears. Similarly, for the bottleneck problem, the vector $d_1^{(0)}$ that initialized the double-sweep algorithm should consist of entries that are smaller than the corresponding optimal entries (since now minimization has been changed to maximization).

The Floyd, Dantzig, and double-sweep algorithms can be extended to even another path problem, the gain problem. Associate a real number, called the *gain factor*, with each arc in the graph. Let the *gain of a path* be defined as the product of the gain factor of each arc in the path. (If an arc is repeated in a path, then repeat its gain factor in the calculation of the path's gain.) The *gain problem* is the problem of finding a path with the largest gain between two vertices.

EXAMPLE 2. The finance officer of a large company has decided to invest the company's surplus funds in bonds that mature within the next five years. A variety of bonds that mature within the next five years are available. Funds released by bonds that mature early can be reinvested in other bonds so long as the funds are ultimately released within five years. How should the finance officer spread his investments over the next five years?

Let the beginning of each investment year be depicted as a vertex. Let each variety of bond be depicted as an arc from the vertex repre representing its date of issue to the vertex depicting its maturity date. Let the gain factor of each arc equal the value at maturity of one dollar invested in the corresponding bond. When presented in graph terms, the finance officer's problem can be viewed as the problem of finding a path from vertex 0 to vertex 5 with the greatest gain.

EXAMPLE 3. A certain amount of pilferage occurs during each segment of a cargo's journey from the factory to the sales outlef. How should the shipment be routed so as to minimize the amount of pilferage?

This problem can be rephrased as a gain problem if we set the gain factor of each arc equal to the fraction of the cargo that is not pilfered during shipment across that arc.

If we replace arc length by arc gain factor, and if we replace addition by multiplication and if we replace minimization by maximization, then the Floyd, Dantzig, and double-sweep algorithms

3.4 Other Shortest Paths

will calculate the paths with the largest gains instead of the shortest lengths. However, just as these shortest path algorithms failed when circuits with negative length were present, these algorithms cannot find the path with the greatest gain when there is a circuit with gain greater than one. This follows since this circuit may be repeated infinitely often to form a (nonsimple) path with infinite gain.

Similarly, if we wanted to find a path between two vertices with the smallest gain, we would retain the minimization operation in the preceding algorithm instead of replacing it with maximization. Again, the algorithms will fail if the graph contains a circuit with gain less than -1. This follows since this circuit could be repeated infinitely often to produce a (nonsimple) path whose gain approaches $-\infty$.

In summary, the Floyd, Dantzig, and double-sweep algorithms can be modified to handle additional problems such as the bottleneck and gains problems. This is accomplished by substituting different operations for the addition and minimization operations in these algorithms and by changing appropriately the arc length matrix.

EXERCISES

1. Use the Dijkstra algorithm to calculate the shortest path from vertex 1 to every other vertex in the graph in Fig. 3.6.

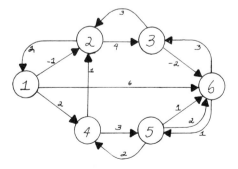

Figure 3.6

2. Suppose that after completing Exercise 1 you discover that there is an arc (4,2) with length 1 missing. Is it possible to salvage these results or must the Dijkstra algorithm be performed over again in its entirety?
3. Calculate the shortest path between every pair of vertices in the graph in Fig. 3.6 using the Floyd algorithm.
4. Calculate the shortest path between every pair of vertices in the graph in Fig. 3.6 using the Dantzig algorithm. Compare these results with the results from the preceding exercise.
5. Calculate the three shortest paths between every pair of vertices in the graph in Fig. 3.6 using the generalized Floyd algorithm. Repeat using the generalized Dantzig algorithm.
6. Suppose you are assigned the task of finding the shortest flight route between all pairs of airports in Europe. However, certain passengers, due to visa restrictions, cannot take flights that stop over in airports in Socialist countries. What is the most efficient way to calculate all the flight routes that are required?
7. Does the arbitrary numbering of the vertices influence the efficiency of the Floyd algorithm? Dantzig algorithm? Double-sweep algorithm? Why?
8. Calculate the three shortest paths from vertex 1 to each other vertex in the graph in Fig. 3.6.
9. Show that a nonoptimal solution may be repeated during two consecutive sweeps of the double-sweep algorithm but may not be repeated for three consecutive sweeps of the double-sweep algorithm.
10. Suppose that the generalized Floyd algorithm was used to calculate the best three bottlenecks between every pair of vertices in a graph. How can this information be used to determine the path corresponding to each bottleneck?
11. In the gains problem, what values should be assigned to the components of the vector d_{ii}^0?
12. The spillage rates between the various tanks in a petroleum refinery are

Exercises

Tank	A	B	C	D
A	0.00	0.13	0.14	0.15
B	0.08	0.00	0.13	0.08
C	0.17	0.12	0.00	0.18
D	0.10	0.06	0.13	0.00

Find the two best ways to ship petroleum from tank C to tank D.

13. Upon terminating the double-sweep algorithm, you discover that one of your arc lengths was incorrect. Under what conditions can you salvage some of your results from the double-sweep algorithm?

14. A pharmaceutical firm wishes to develop a new product within the next 12 months to compete with a similar product recently marketed by their chief competitors.

    The development of a new product requires four stages each of which can be performed at the slow, normal or fast pace. The times and costs of these are

	Theoretical research[a]	Laboratory experiments[a]	Government approval[a]	Marketing[a]
Slow	5,5	3,6	6,1	5,8
Normal	4,7	2,8	4,1	4,10
Fast	2,10	1,12	2,3	3,15

[a]Time in months, cost in thousands.

What is the best way for the firm to have the new product ready within 12 months without spending more than $25,000? What is the second best way to achieve this same goal?

15. A hotel manager must make reservations for the bridal suite for the coming month. He has received a variety of reservation requests for various combinations of arrival and departure days. Each reservation would earn a different amount of revenue for the hotel due to a variety of rates for students, employees,

airline personnel, etc. How can the Dijkstra algorithm be used to find the best way to schedule the bridal suite with maximum profits to the hotel?

(Hint: Represent each reservation request by an arc joining its arrival date to its departure date. The resulting graph will contain no circuits. A variation of the Dijkstra algorithm can be applied to this situation.)

REFERENCES

Dantzig, G. B., 1967. All Shortest Routes in a Graph, *Theory of Graphs, International Symposium, Rome, 1966,* Gordon and Breach, New York, pp. 91-92.

Dijkstra, E. W., 1959. A Note on Two Problems in Connexion with Graphs, *Numer. Math., 1,* pp. 269-271.

Dreyfus, S. E., 1969. An Appraisal of Some Shortest Path Algorithms, *Operations Research, 17,* pp. 395-412.

Floyd, R. W., 1962. Algorithm 97, Shortest Path, *Comm. ACM, 5,* p. 345.

Ford, L. R., 1956. Network Flow Theory, *Rand Corporation Report,* p. 923.

Karp, R. M., 1975. On the Computational Complexity of Combinatorial Problems, *Networks, 5,* pp. 45-68.

Knuth, D. E., 1973. *The Art of Computer Programming, vol. 3, Sorting and Searching,* Addison-Wesley, Reading.

Lawler, E. L., 1971. The Complexity of Combinatorial Computations: A Survey, *Proc. 1971 Polytechnic Institute of Brooklyn Symposium on Computers and Automata.*

Minieka, E. T., 1974. On Computing Sets of Shortest Paths in a Graph, *Comm. ACM, 17,* pp. 351-353.

Minieka, E. T., and D. R. Shier, 1973. A Note on an Algebra for the k Best Routes in a Network, *J. IMA, 11,* pp. 145-149.

Shier, D. R., 1976. Iterative Methods for Determining the k Shortest Paths in a Network, *Networks, 6,* pp. 205-230.

Shier, D. R., 1974. Computational Experience with an Algorithm for Finding the k Shortest Paths in a Network, *J. Res. Natl. Bur. Std., 78B* (July-September), pp. 139-165.

Williams, T. A., and G. P. White, 1973. A Note on Yen's Algorithms for Finding the Length of All Shortest Paths in N-Node Nonnegative Distance Networks, *J. ACM, 20,* No. 3, pp. 389-390.

References

J. Y. Yen, 1970. An Algorithm for Finding Shortest Routes from All Source Nodes to a Given Destination in General Networks, *Q. Appl. Math., 27*, pp. 526-530.

J. Y. Yen, 1973. Finding the Lengths of All Shortest Paths in N-Node Nonnegative Distance Complete Networks Using $1/2N^3$ Additions and N^3 Comparisons, *J. ACM, 19*, No. 3, pp. 423-424.

Chapter 4

FLOW ALGORITHMS

4.1 INTRODUCTION

Loosely speaking, a *flow* is a way of sending objects from one place to another. For example, the shipment of finished goods from a manufacturer to a distributor, the movement of people from their homes to places of employment, or the delivery of letters from their point of posting to their destinations can all be regarded as flows.

Despite the variety of flow situations, there are several fairly common problems that arise in flow situations. For example, one might wish to maximize the amount transported from one place to another by the delivery system, or one might wish to determine the least cost way to send a given number of objects from one place to another via the system, or one might wish to determine the quickest way to deliver a shipment through the system.

In this chapter, a flow will be defined in terms of a graph, and various flow problems, like those suggested above, will be presented. If the original flow problem can be modeled in terms of a graph flow problem with a high degree of accuracy, then results and algorithms for graph flows can be applied to the original problem. Needless to say, such results are usually no better than the graph model used to obtain them.

Loosely speaking, a flow on a graph is a way of sending objects from one vertex to another by traveling along the arcs in their directions. The vertex from which the objects start their travels is called the *source* and is usually denoted by s. The vertex at which the objects end their travels is called the *sink* and is usually denoted by t. The objects that travel or flow from the source to the sink are called *flow units* or *units*. As mentioned above, the flow units can be finished goods, people, letters, or almost anything.

If the number of flow units that can travel across arc (x,y) is limited, then arc (x,y) is called a *capacitated* arc. We shall denote the maximum *capacity* of arc (x,y) by c(x,y), and we shall denote the *cost* of sending one unit across arc (x,y) by a(x,y). A *network* is a graph in which each arc has a capacity associated with it.

EXAMPLE 1. A wartime convoy of twelve trucks carrying military supplies must be dispatched from the depot (source) to the troops (sink). The roads connecting the supply depot to the troops can be represented as a graph in which the edges correspond to unprotected roads and the vertices correspond to road junctions. For security reasons, a limit is placed on the number of trucks that may travel each road (arc capacity). From previous experience, the dispatcher knows the amount of pilferage he can expect from a truck along each road segment (arc cost).

The problem of finding the best routes for each of the 12 supply trucks can be rephrased as the problem of sending 12 flow units from the source to the sink in the network corresponding to the roads so that no arc's capacity is exceeded and so that total expected amount of pilferage is minimized.

Suppose we are given a graph in which a certain number of units are traveling from the source to the sink, and suppose that the route that each unit takes is also known to us. Let the number of units that traverse arc (x,y) be called the flow in arc (x,y) and be denoted by f(x,y). Of course, $0 \leq f(x,y) \leq c(x,y)$. The arcs of the graph can be classified into three categories:

N, the set of arcs that cannot allow any increase or decrease in their flows

I, the set of arcs whose flow can be increased

R, the set of arcs whose flow can be reduced.

4.1 Introduction

For example, an arc with zero capacity or a prohibitive traverse cost would belong to set N. Arcs with unused capacity would belong to set I, and arcs with flow on them already would belong to set R. Arcs in set I are called *increasable*; arcs in set R are called *reducable*. Clearly, each arc must belong to at least one of these three sets N, I, R. Possibly, an arc could belong to both sets I and R; this occurs when the arc already carries flow units but can have its flow increased or reduced. Such an arc is called *intermediate*.

Let i(x,y) denote the maximum amount by which the flow in arc (x,y) can be increased. Similarly, let r(x,y) denote the maximum amount by which the flow in arc (x,y) can be decreased. Thus, i(x,y) = c(x,y) - f(x,y) and r(x,y) = f(x,y).

Suppose we wanted to send some additional units from s to t. There are several ways to accomplish this (provided, of course, that it is possible to send more from s to t). First, if you could find a path P from s to t consisting entirely of increasable arcs, then additional flow could be sent from s to t along path P (see Fig. 4.1). How many additional flow units could be sent from s to t along path P? Since i(x,y) denotes the maximum amount by which the flow in arc (x,y) can be increased, then at most

$$\min_{(x,y) \in P} \{i(x,y)\}$$

additional flow units can be sent from s to t. For the path P in Fig. 4.1 one additional flow unit can be sent from s to t since

$$\min\{i(s,a), i(a,b), i(b,t)\} = \min\{3,2,1\} = 1$$

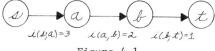

Figure 4.1

Flow Augmenting Path with Only Forward Arcs

Secondly, if we could find a path P from t to s consisting entirely of decreasable arcs, then flow could be decreased in each arc in this path resulting in less flow from t to s and consequently a greater net flow from s to t (see Fig. 4.2). What is the maximum amount by which the flow in this path could be decreased? Since each arc (x,y) can have its flow decreased by at most r(x,y), the maximum flow decrease in path P is

min{r(x,y)}
(x,y) ∈ P

For the path in Fig. 4.2 one flow unit from t to s can be reversed since

min{r(t,b), r(b,a), r(a,s)} = min{1,2,1} = 1

Is it possible to find yet another way to increase the net flow from s to t? Yes, we could combine the two methods shown above: namely, we could find a chain from s to t with the following properties:

1. The arcs in the direction of s to t, called *forward* arcs, are all members of I
2. The arcs in the direction of t to s, called *backward* arcs, are all members of R.

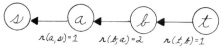

Figure 4.2

Flow Augmenting Path with Only Backward Arcs

For example, consider chain C from s to t in Fig. 4.3. The forward arcs are (s,a), (a,b) and (d,t); the backward arcs are (c,b) and (d,c). If each forward arc belongs to I and each backward arc belongs to R, then additional flow can be sent from s to t along this chain by advancing along the forward arcs which are increasable

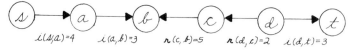

Figure 4.3

Flow Augmenting Chain with
Forward and Backward Arcs

4.1 Introduction

and reversing along the backward arcs which are decreasable. The maximum amount of additional flow that can be sent along such a chain from s to t is the minimum of the following two quantities:

$\min\{i(x,y): (x,y)$ is a forward arc$\}$
$\min\{r(x,y): (x,y)$ is a backward arc$\}$

The minimum of these two quantities is called the *maximum flow augmentation of the chain*. In Fig. 4.3, the amounts by which the forward arcs can be increased are $i(s,a) = 4$, $i(a,b) = 3$ and $i(d,t) = 3$, the minimum of which is 3. The amounts by which the backward arcs can be decreased are $r(c,b) = 5$ and $r(d,c) = 2$, the minimum of which is 2. Hence, the maximum flow augmentation of this chain equals min $(3,2) = 2$. Thus, 2 additional flow units are sent from s to t along this chain by increasing the flow in each of the three forward arcs by 2 units and decreasing the flow in each of the two backward arcs by 2 units.

Any chain from s to t, like the three different types discussed above, along which additional flow units can be sent is called a *flow augmenting chain*. How can we determine if there exists a flow augmenting chain from s to t? This can be done quite simply using the following algorithm called the *flow augmenting algorithm*. The essential idea of this algorithm is to grow out of source s a tree of colored arcs along which additional flow units can be sent from s. The algorithm either colors sink t in which case the unique chain in the colored tree from s to t is a flow augmenting chain from s to t or the algorithm does not color sink t in which case no flow augmenting chain exists from s to t for the current classification of the arcs into sets R, I, N.

Flow Augmenting Algorithm

Step 1: Determine which arcs belong to sets N, I, and R. The arcs in set N can be ignored by the algorithm since there can be no flow changes in them. Color vertex s.

Step 2: Color the arcs and vertices according to the following rules until vertex t has been colored or until no further coloring is possible:

If vertex x is colored and vertex y is not colored, then vertex y can be colored and the arc (x,y) can be colored if either

(a) vertex g and arc (x,y) can be colored if arc $(x,y) \in I$, or

(b) vertex y and arc (y,x) can be colored if arc $(y,x) \in R$

If vertex t has been colored, then there exists a unique chain of colored arcs from s to t. This chain is a flow augmenting chain. Otherwise, if t remains uncolored after the algorithm terminates, then no flow augmenting chain exists from s to t. Stop.

Proof of the flow augmenting algorithm: To prove that the algorithm locates a flow augmenting chain if one exists, three facts must be demonstrated:

(1) If t is colored by the algorithm, then there does in fact exist a flow augmenting chain from s to t

(2) If t cannot be colored by the algorithm, then there does not exist any flow augmenting chain from s to t.

(3) The algorithm terminates after a finite number of steps.

Proof of (1): Due to the restrictions of the algorithm's coloring procedure, an arc is colored only if one of its endpoints is colored and one of its endpoints is not colored. Hence, the algorithm can never color an arc with two already colored endpoints, and thus ths colored arcs can never form a cycle. Since only vertex s is colored initially, the colored arcs must form a tree that contains vertex s. Thus, if vertex x is colored by the algorithm, there must exist a unique chain of colored arcs from s to x. Specifically, if t is colored, there exists a unique chain of colored arcs from s to t. This chain must be a flow augmenting chain since the coloring procedure insures that the chain's forward arcs are increasable and the chain's backward arcs are decreasable.

4.1 Introduction

Proof of (2): If there exists a flow augmenting chain C from s to t, then there exists at least one flow augmenting chain from s to each vertex in chain C. Thus, each vertex in chain C can be colored by the algorithm, and hence vertex t must also be colored. Conversely, if t cannot be colored by the algorithm, then no flow augmenting chain from s to t exists.

Proof of (3): Finally, the algorithm must terminate after a finite number of colorings since there are only a finite number of vertices and arcs that can be colored and no vertex or arc can be colored more than once. Q.E.D.

EXAMPLE 2 *(Flow Augmenting Algorithm)*. We shall apply the flow augmenting algorithm to find a flow augmenting chain from s to t in Fig. 4.4. The letters next to each arc indicate whether it is increasable and/or decreasable. (In this example, we shall color every possible vertex adjacent to a given colored vertex before attempting to color from another vertex.)

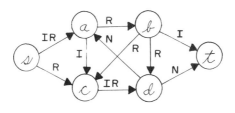

$i(s,a) = 4 \quad r(s,c) = 1$
$i(a,c) = 3 \quad r(a,b) = 7$
$i(b,t) = 2 \quad r(c,d) = 2$
$i(c,d) = 5 \quad r(b,d) = 3$
$d(s,a) = 6 \quad r(b,c) = 2$

Figure 4.4

Example of a Flow Augmenting Algorithm

Initially, vertex s is colored. From vertex s, vertex a and arc (s,a) can be colored since (s,a) ∈ I. Vertex c and arc (s,c) cannot be colored from vertex s since (s,c) ∉ I. This completes the coloring out of vertex s.

Next, we shall consider coloring out of vertex a. Vertex b and arc (a,b) cannot be colored since (a,b) ∉ I. Vertex c and arc (a,c) can be colored since (a,c) ∈ I. Vertex d and arc (d,a) cannot be colored from vertex a since (d,a) ∉ R. This completes the coloring out of vertex a.

Next, we shall consider coloring out of vertex c. Since vertices s and a are already colored, they can be ignored. Vertex b and arc (b,c) can be colored since (b,c) ∈ R. Also, vertex d and arc (c,d) can be colored since arc (c,d) ∈ I. This completes the coloring out of vertex c.

Next, we shall consider coloring out of vertex b. Vertices a, c and d have already been colored and can be ignored. Vertex t and arc (b,t) can be colored since arc (b,t) ∈ I.

The algorithm stops coloring since t has been colored. The colored arcs and vertices are shown in Fig. 4.5. The flow augmenting chain from s to t is

(s,a), (a,c), (b,c), (b,t)

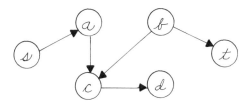

Figure 4.5
Colored Tree

The maximum flow augmentation is

min{i(s,a), i(a,c), r(b,c), i(b,t)} = min{4,3,2,2} = 2

Forward arcs (s,a), (a,c), and (b,t) can have their flows increased by 2 units; backward arc (b,c) can have its flow decreased by 2 units which implies that 2 units formerly flowing through (b,c) would be rerouted along arc (b,t) and be replaced at vertex c by two additional flow units coming from s via arcs (s,a) and (a,c).

4.1 Introduction

As can be seen in Example 2, the flow augmenting algorithm does not specify completely which arc and vertex is to be colored next. The algorithm could color as much as possible from a given vertex (as it did in the above example) or it could color out of the most recently colored vertex. The most efficient way to color is a matter of some research and often depends upon the larger problem in which the flow augmenting algorithm is being used as a subalgorithm.

Once the algorithm has located a flow augmenting chain, additional flow units not exceeding the maximum flow augmentation can be sent from s to t along the flow augmenting chain by increasing the flow in its increasable arcs and decreasing the flow in its decreasable arcs by the quantity that is being sent.

4.2 MAXIMUM FLOW ALGORITHM

The *maximum flow problem* is simply the problem of finding a way to send the maximum number of flow units from the source to the sink in a capacitated graph so that no arc capacity is violated.

EXAMPLE 1. A travel agent must arrange for the flights of a group of ten tourists from Chicago airport to Istanbul airport on a certain day. On that day, there are 7 seats left on the Chicago-Istanbul direct flight; there are 5 seats left on the Chicago-Paris flight, and there are 4 seats left on the connecting Paris-Istanbul flight. What should the agent do?

The travel agents problem can be posed as a maximum flow problem. Construct a network in which each arc represents a flight. There are three arcs (Chicago, Istanbul), (Chicago, Paris) and (Paris, Istanbul). Assign a capacity to each arc that equals the number of available seats on the corresponding flight, that is, 7,5,4, respectively. If this network admits a flow from Chicago (source) to Istanbul (sink) of 10 or more units without violating any arc capacities, then the travel agent can send the entire group on the selected day.

As mentioned in Sec. 4.1, a flow is any shipment from s to t. As before, let the number of units traveling through arc (x,y) be denoted by f(x,y). In any flow from s to t, the number of units that leave each vertex x (x ≠ s, x ≠ t) must equal the number of units that enter vertex x, that is,

$$\sum_{y \in X} f(x,y) = \sum_{y \in X} f(y,x) = 0 \quad \text{(for all } x \neq s, s \neq t) \quad (1)$$

(Recall that X is the set of all vertices.) Furthermore, the total number of flow units that travel across arc (x,y) must not exceed c(x,y) the capacity of arc (x,y), that is,

$$0 \leq f(x,y) \leq c(x,y) \quad \text{(for all } (x,y) \in A) \quad (2)$$

(Recall that A is the set of all arcs.)

Also, the net number v of flow units that leave the source must equal the net number of flow units that enter the sink, that is,

$$\sum_{y \in X} f(x,y) - \sum_{y \in X} f(y,s) = v \quad (3)$$

$$\sum_{y \in X} f(y,t) - \sum_{y \in X} f(t,y) = v \quad (4)$$

Every flow from s to t must satisfy these four conditions. Moreover, if a set of values f(x,y), (x,y) ∈ A, can be found that satisfy these four conditions, then these values correspond to a flow from s to t. This can be shown by tracing the path that each flow unit takes on its journey from s to t. The details are left to the reader.

Thus, a set of values f(x,y), (x,y) ∈ A, is a flow if, and only if, it satisfies relations (1)-(4).

The maximum flow problem is simply to find the maximum value of v for which a flow exists, i.e., for which relations (1)-(4) are satisfied. The reader familiar with linear programming will note that maximizing v subject to the restrictions of relations (1)-(4) is a linear programming problem. Since the maximum flow problem is a linear programming problem, it could be solved by the simplex algorithm for linear programming problems. However, using the simplex algorithm for this problem is like killing a mouse with a cannon. A more elegant and far more intuitive approach is available. This is the Ford and Fulkerson maximum flow algorithm (Ford and Fulkerson, 1962, p. 4), henceforth called the *maximum flow algorithm*.

4.2 Maximum Flow Algorithm

The idea underlying the maximum flow algorithm is quite simple: Start with any flow from s to t and look for a flow augmenting chain using the flow augmenting algorithm. If a flow augmenting chain from s to t is found, then send as many flow units as possible along this chain. Then, start again to look for another flow augmenting chain, etc. If no flow augmenting chain is found, then stop because the current flow from s to t is shown to be a maximum flow from s to t.

With these ideas in mind we can state formally the maximum flow algorithm for finding a maximum flow from s to t in a *network*, a graph with arc capacities.

Maximum Flow Algorithm

Step 1: Let s denote the source vertex, and let t denote the sink vertex. Select any initial flow from s to t, i.e., any set of values for $f(x,y)$ that satisfy relations (1)-(4). If no such initial flow is known, use as the initial flow $f(x,y) = 0$ for all (x,y).

Step 2: If $f(x,y) < c(x,y)$, let $i(x,y) = c(x,y) - f(x,y)$ and $(x,y) \in I$. If $f(x,y) > 0$, let $r(x,y) = f(x,y)$ and $(x,y) \in R$.

Step 3: For sets I and R defined in Step 2, perform the flow augmenting algorithm. If no flow augmenting chain is discovered by the flow augmenting algorithm, stop; the current flow is a maximum flow. Otherwise, make the maximum possible flow augmentation along the flow augmenting chain discovered by the flow augmenting algorithm. Return to Step 2.

Proof of the maximum flow algorithm: To show that the maximum flow algorithm constructs a maximum flow from s to t, we must show that

(1) The algorithm constructs a flow
(2) This flow is a maximum flow
(3) The algorithm terminates after a finite number of steps

Proof of (1): To show that the algorithm constructs a flow, we need only note that the algorithm is initialized in Step 1 with a flow and that during all steps, the algorithm maintains a flow since each flow augmentation maintains a flow. Hence, the algorithm must terminate with a flow.

Proof of (2): Recall from Chap. 1 that a cut is defined as any set of arcs whose removal disconnects the graph. Also recall that a simple cut is a cut that contains no other cut as a proper subset. In Fig. 4.6, arcs (s,a) and (s,b) form a cut since their removal would disconnect the graph into two components. This cut is a simple cut since neither arc is a cut by itself.

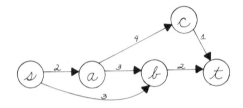

Figure 4.6
Example of a Maximum Flow Algorithm

Consider any simple cut that separates the source into one component and the sink into another component. Let the component containing the source be denoted by X_s and let the component containing the sink be denoted by X_t. Each arc in this cut must either (a) have its head in X_s and its tail in X_t, or (b) have its head in X_t and its tail in X_s. Let the sum of the capacities of the arcs of type (b) be called the *capacity of the cut*. Since the cut separates s from t, clearly, it is impossible to send more flow units from s to t than the capacity of this cut since every flow unit must cross the cut. In general, the maximum flow possible from s to t must surely be less than or equal to the smallest cut capacity of a cut separating s from t. Simply,

max flow $\leq$ min cut

In order to demonstrate that the maximum flow algorithm has indeed produced a maximum flow we need only produce a cut separating s from t whose capacity equals the value of the terminal flow produced by the algorithm. This is done as follows:

4.2 Maximum Flow Algorithm

The algorithm terminates when no flow augmenting chain can be found from s to t. After the final application of the flow augmenting algorithm, vertex t could not be colored. Consider the cut consisting of all arcs that upon termination of the last application of the flow augmenting algorithm have one endpoint colored and the other endpoing not colored. Since s is colored and t is not colored, this cut separates s from t. The capacity of this cut is the sum of the capacities of the arcs with colored tails and uncolored heads.

Upon termination of the maximum flow algorithm, each arc with uncolored head and colored tail carries a flow equal to its capacity; otherwise, the head of this arc could be colored by the flow augmenting algorithm. Upon termination of the maximum flow algorithm, each arc with uncolored tail and colored head carries no flow units; otherwise, the tail of this arc could be colored by the flow augmenting algorithm.

Clearly, every flow unit must traverse an arc of this cut at least once. Since all arcs from the sink side to the source side of the cut carry no flow, no flow unit can traverse this cut more than once. Thus, the total flow from s to t equals the capacity of this cut becuase every arc from the source side to the sink side carries a capacity flow.

Proof of (3): To show that the algorithm terminates in a finite number of steps requires the assumption that all arc capacities are integers and all initial flow values are integers. Practically speaking, this is not a drastic assumption since most arc capacities can usually be rounded off to an integer without affecting the underlying physical problem.

The only possible way for the maximum flow algorithm not to terminate in a finite number of steps would be for the algorithm to encounter an infinite number of flow augmenting chains. However, each time a flow augmenting chain is found, the total flow v from s to t is increased by a positive integer because all flow values and arc capacities are always integers. Since v is bounded above by the capacity of any cut separating s from t, there cannot be an infinite number of flow augmentations. Q.E.D.

EXAMPLE 2. Consider the graph in Fig. 4.6. The number adjacent to each arc denotes the arc capacity.

Step 1: Initialize the algorithm with the zero flow, that is, $f(x,y) = 0$ for all arcs (x,y).

Step 2: Since $f(x,y) < c(x,y)$ for all arcs (x,y), every arc is a member of I, and $i(x,y) = c(x,y) - f(x,y) = c(x,y)$. Furthermore, since $f(x,y) = 0$ for all arcs (x,y), no arc is a member of R.

Step 3: The flow augmenting algorithm is now used to find a flow augmenting chain from s to t. Obviously, by inspection, there are several flow augmenting chains from s to t. Suppose that the flow augmenting chain (s,a), (a,b), (b,t) is generated. The maximum flow augmentation along this chain is $\min\{i(s,a), i(a,b), i(b,t)\} = \min\{2,3,2\} = 2$. So, the flow in each of the three arcs in this chain is increased by two units: $f(s,a) = 2$, $f(a,b) = 2$, $f(s,b) = 0$, $f(b,t) = 2$, $f(a,c) = 0$, and $f(c,t) = 0$. Return to Step 2.

Step 2: For the new flow generated above,

$f(s,a) = 2 = c(s,a)$ $(s,a) \notin I$ $(s,a) \in R, r(s,a) = 2$
$f(a,b) = 2 < c(a,b)$ $(a,b) \in I, i(a,b) = 1$ $(a,b) \in R, r(a,b) = 2$
$f(b,t) = 2 = c(b,t)$ $(b,t) \notin I$ $(b,t) \in R, r(b,t) = 2$
$f(s,b) = 0 < c(s,b)$ $(s,b) \in I, i(s,b) = 3$ $(s,b) \notin R$
$f(a,c) = 0 < c(a,c)$ $(a,c) \in I, i(a,c) = 4$ $(a,c) \notin R$
$f(c,t) = 0 < c(c,t)$ $(c,t) \in I, i(c,t) = 1$ $(c,t) \notin R$

Step 3: The flow augmenting algorithm is applied again to find another flow augmenting chain. The unique flow augmenting chain that it will generate is (s,b), (a,b), (a,c), (c,t). The maximum flow augmentation possible along this chain is

$\min\{i(s,b), r(a,b), i(a,c), i(c,t)\} = \min\{3,2,4,1\} = 1$

Thus, one additional unit of flow is sent along this chain from s to t. The flow in each forward arc (s,b), (a,c), and (c,t) of the chain is increased by one unit, and the flow in the backward arc (a,b) of the chain is decreased by one unit. The resulting flow is $f(s,a) = 2$, $f(a,b) = 1$, $f(b,t) = 2$, $f(s,b) = 1$, $f(a,c) = 1$, $f(c,t) = 1$. The flow units now travel the following routes:

4.2 Maximum Flow Algorithm

1 unit travels from s to t via (s,a), (a,b), (b,t)
1 unit travels from s to t via (s,b), (b,t)
1 unit travels from s to t via (s,a), (a,c), (c,t).

Return to Step 2.

For the flow generated above,

$f(s,a) = 2 = c(s,a)$	$(s,a) \notin I$	$(s,a) \in R, \; r(s,a) = 2,$
$f(a,b) = 1 < c(a,b)$	$(a,b) \in I, \; i(a,b) = 2,$	$(a,b) \in R, \; r(a,b) = 1,$
$f(b,t) = 2 = c(b,t)$	$(b,t) \notin I,$	$(b,t) \in R, \; r(b,t) = 2,$
$f(s,b) = 1 < c(s,b)$	$(s,b) \in I, \; i(s,b) = 2,$	$(s,b) \in R, \; r(s,b) = 1,$
$f(a,c) = 1 < c(a,c)$	$(a,c) \in I, \; i(a,c) = 3,$	$(a,c) \in R, \; r(a,c) = 1,$
$f(c,t) = 1 = c(c,t)$	$(c,t) \notin I,$	$(c,t) \in R, \; r(c,t) = 1.$

Applying the flow augmenting algorithm to the graph with the above flow as the initial flow, we find that no additional flow augmenting chains exist. The flow augmenting algorithm could color vertices s, b, a, c (in that order), but not vertex t. Since no more flow augmenting chains can be found, the maximum flow algorithm terminates. The terminal flow (described above) is a maximum flow from s to t and hence at most three units can be sent from s to t.

Note that the cut formed by the arcs with one endpoint colored and one endpoint uncolored after the last application of the flow augmenting algorithm is (b,t), (c,t). The capacity of this cut is $c(b,t) + c(c,t) = 2 + 1 = 3$, which is the number of units sent from s to t by the maximum flow.

The remainder of this section describes two important modifications of the maximum flow algorithm. The first modification shows how to insure that the maximum flow algorithm will terminate in a finite number of steps when we cannot assume that all flows and arc capacities are integers. The second modification describes how to solve the maximum flow problem for graphs with more than one source vertex and more than one sink vertex.

Finite Termination Modification

The proof of the maximum flow algorithm required that all initial arc flow values were integers and that all arc capacities were also integers. If some arc capacities are not integers, there is no guarantee that the maximum flow algorithm will terminate finitely. An example of a graph with noninteger arc capacities for which the maximum flow algorithm requires an infinite number of flow augmentations can be found in Ford and Fulkerson (1962). Fortunately, Johnson (1966) and Edmonds and Karp (1972) have provided two different ways to modify the maximum flow algorithm to ensure that it terminates after only a finite number of flow augmentations. The modification due to Edmonds and Karp is presented now.

The flow augmenting algorithm may have a choice of arcs to color next. The finite termination modification specifies the next arc to be colored as follows: Number the vertices as they are colored. (Of course, vertex s will receive number one.) First, color all possible arcs incident to vertex number one. Next, color all possible arcs incident to vertex number two, etc.

Note that if the coloring is performed as described above, then the chain of colored arcs connecting any vertex to the source will contain as few arcs as possible. Hence, each flow augmenting chain generated by this modified coloring method will contain as few arcs as possible.

Arc (x,y) is called a *bottleneck* arc whenever arc (x,y) limits the amount of the flow augmentation. If arc (x,y) is a bottleneck arc, either f(x,y) increases to c(x,y) or reduces to zero.

Suppose that arc (x,y) is a bottleneck arc in both flow augmenting chains C_1 and C_2 but not in any flow augmenting chain occurring between C_1 and C_2. Without loss of generality, suppose that f(x,y) = c(x,y) after the flow augmentation along chain C_1. Thus, (x,y) must be a forward arc in C_1 and a backward arc in C_2. For i = 1, 2, denote the number of arcs from vertex m to vertex n in flow augmenting chain C_i by $C_i(m, n)$. From the observation that the modified coloring method always colors the shortest possible flow augmenting chain from s to any vertex, it follows that

4.2 Maximum Flow Algorithm

$$C_1(s,y) \leq C_2(s,y)$$
$$C_1(x,t) \leq C_2(x,t)$$
$$C_1(s,t) = C_1(s,y) + C_1(x,t) - 1$$
$$\leq C_2(s,y) + C_2(x,t) - 1$$
$$= C_2(s,t) - 2.$$

Thus, $C_1(s,t) \leq C_2(s,t) - 2$, and each time arc (x,y) is a bottleneck arc, the minimum number of arcs in a shortest flow augmenting chain has increased by at least two. Since no flow augmenting chain from s to t can contain more than $n - 1$ arcs (recall that n is the number of vertices in the graph), it follows that arc (x,y) cannot be a bottleneck arc more than $n/2$ times. Since each flow augmenting chain has at least one bottleneck arc, there can be at most $n|A|/2$ flow augmentations.

A similar result follows if the flow in arc (x,y) is reduced to zero by the flow augmentation in chain C_1.

EXAMPLE 3. Let us use the modified coloring method to find a maximum flow in the graph shown in Fig. 4.6. Suppose that initially each arc carries no flow.

First, vertex s is colored, and vertex s receives number one. Examining the uncolored arcs incident to vertex s, we find that arc (s,a) and vertex a can be colored. Vertex a receives number two. Next, arc (s,b) and vertex b can be colored. Vertex b receives number three. This completes the coloring of arcs incident to vertex s.

Since vertex a received number 2, we now examine the uncolored arcs incident to vertex a. This results in the coloring of arc (a,c) and vertex c. Vertex c receives number four.

Since vertex b receives number three, we next examine the uncolored arcs incident to vertex b. This results in the coloring of arc (b,t) and vertex t. The flow augmenting chain (s,b), (b,t) has been found. A flow of 2 units is sent from s to t along this chain.

All previous coloring and numbering is erased, and the coloring process is repeated. Vertex s is colored and receives number one. Examining the uncolored arcs incident to vertex s results in the coloring of arc (s,a) and vertex a and in the coloring of arc (s,b) and vertex b. Vertices a and b are respectively numbered two and three.

Next, the uncolored arcs incident to vertex a are examined. This results in the coloring of arc (a,c) and vertex c. Vertex c receives number four.

Next, the uncolored arcs incident to vertex b are examined. This results in no further coloring. Next, the uncolored arcs indident to vertex c are examined. This results in the coloring of arc (c,t) and vertex t. A flow augmenting chain (s,a), (a,c), (c,t) has been discovered. One additional flow unit is sent from s to t along this chain. It is left to the reader to verify that no additional flow is possible.

Modification for Several Sources and Sinks

Lastly, let us consider a graph in which there are possibly more than one source vertices and more than one sink vertices. Can this situation be accommodated by the maximum flow algorithm which works with only one source and one sink? Yes, simply create a new source vertex S, called the *supersource*, and a new sink vertex T, called the *supersink*. Join the supersource S to each original source s_1, s_2, ..., by an arc (S,s_1), (S,s_2), ... with infinite capacity. Join each original sink t_1, t_2, ... to the supersink T by an arc (t_1,T), (t_2,T), ..., with infinite capacity.

Clearly, any flow on the new, enlarged graph from S to T corresponds to a flow on the original graph from the original sources to the original sinks, and viceversa. Moreover, a maximum flow in the enlarged graph corresponds to a maximum flow in the original graph. Thus, the maximum flow algorithm can be applied to the enlarged graph, and the maximum flow generated by the algorithm yields a maximum flow in the original graph. See Fig. 4.7.

4.3 Minimum Cost Flow Algorithm

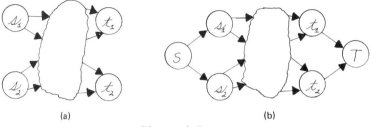

(a)   (b)

Figure 4.7

Graph with Several Sources and Sinks:

(a) Original Graph

(b) Expanded Graph

4.3 MINIMUM COST FLOW ALGORITHM

In the preceding section, we considered the maximum flow problem, i.e., how to send the maximum possible number of units from the source to the sink in a capacitated graph. In this section, we shall consider the problem of how to send with minimum cost a given number v of flow units from the source to the sink in a capacitated graph in which there is a cost per unit associated with using each arc.

EXAMPLE 1. A manufacturer can select from a variety of routes to ship finished products from his factory to his warehouse. He incurs a different cost per pound depending upon the shipping route or routes selected. Each route can accommodate only a limited total weight. What is the least cost way for the manufacturer to ship all his finished products to his warehouse?

Represent the factory by a vertex s and represent the warehouse by another vertex t. Let each intersection of two or more routes be represented by a vertex, and let each uninterrupted route segment be represented by an arc between the appropriate vertices. Let the capacity of each arc equal the maximum weight that can be accommodated by the corresponding route setment, and let each arc cost equal the cost per pound for using the corresponding route segment.

The manufacturer's problem can now be viewed as the problem of finding on this graph a minimum cost flow from s to t.

EXAMPLE 2. Suppose that a travel agent has been contacted by a group of 75 people who wish to fly tomorrow (either separately or together) from Springfield to Istanbul. What is the least cost way he can route all of them from Springfield to Istanbul tomorrow?

The travel agent's problem can be rephrased as a minimum cost flow problem on the graph whose vertices are the various airports between Springfield and Istanbul and whose arcs represent the various flights between these airports tomorrow. The capacity of each arc equals the number of seats available tomorrow on the corresponding flight, and the cost of each arc equals the cost of one seat tomorrow on the corresponding flight.

The algorithm presented here to solve the minimum cost flow problem is due to Ford and Fulkerson (1962, p. 113). This algorithm, appropriately called the *minimum cost flow algorithm*, is a generalization of the maximum flow algorithm of Sec. 3.2.

Let $a(x,y)$ denote the cost of sending one flow unit along arc (x,y). Initially, we shall assume that each $a(x,y)$ is a positive integer. This assumption is not very restrictive since costs are usually expressed in dollars and cents which are positive integers. (At the end of this section, we shall present a method to modify the minimum cost flow algorithm so that noninteger costs can be accommodated.)

As before, let $f(x,y)$ denote the number of flow units that travel across arc (x,y). Of course, $f(x,y) \geq 0$. Let v denote the number of units to be sent from source to sink.

The minimum cost flow problem can be expressed as follows:

$$\min\{ \sum_{(x,y)} a(x,y) \, f(x,y) \} \tag{5}$$

such that

$$\sum_y [f(s,y) - f(y,s)] = v \tag{6}$$

$$\sum_y [f(x,y) - f(y,x)] = 0 \quad \text{(for all } x \neq s, \, x \neq t\text{)} \tag{7}$$

$$\sum_y [f(t,y) - f(y,t)] = -v \tag{8}$$

$$0 \leq f(x,y) \leq c(x,y) \quad \text{[for all } (x,y)\text{]} \tag{9}$$

4.3 Minimum Cost Flow Algorithm

Expression (5) represents the total cost of a flow. Equation (6) states that the net flow out of source s must equal v. Equation (7) states that the net flow out of any vertex x that is neither the source nor the sink must equal zero. Equation (8) states that the net flow out of the sink t must equal -v. Condition (9) requires that the flow in each arc take a value between zero and the arc's capacity.

Like the maximum flow problem, the minimum cost flow problem is a linear programming problem. (In fact, the maximum flow problem can be viewed as a minimum cost flow problem in which all arc costs are zero and v is the value of the maximum possible flow.)

Suppose that expression (5) were replaced by

$$\max\{pv - \sum_{(x,y)} a(x,y) f(x,y)\} \qquad (10)$$

where p is any large number, e.g., larger than the maximum total cost a unit could incur by traveling from s to t. If p is interpreted as the profit received for each unit sent from s to t, then the expression in (10) can be interpreted as the best possible net profit after shipping costs have been deducted. From this interpretation, it follows that any flow that maximizes (10) will also minimize (5), and vice-versa.

The minimum cost flow algorithm first sends as many units as possible from s to t that incur a total cost of 0 each for the entire journey from s to t. Next, the minimum cost flow algorithm sends as many units as possible from s to t that incur a total incremental cost of 1 each for the entire journey from s to t, etc. The algorithm stops when a total of v units have been sent from s to t, or when no more units can be sent from s to t, whichever happens first. In other words, the algorithm solves the problem given by (6) - (10) first for p = 0, then for p = 1, then for p = 2, etc.

Suppose that as many units as possible with a total incremental cost of p - 1 or less have been sent by the algorithm from s to t. How does the algorithm determine how to send flow units from s to t with a total incremental cost of p each? To do this, the algorithm must locate a flow augmenting chain from s to t with the property that the total incremental cost of sending a unit "along this chain" equals p. To clarify this idea, consider the graph in Fig. 4.8. The least expensive way to send a flow unit from s to t in this graph is along the path (s,a), (a,b), (b,t). This flow unit incurs a total cost of $2 + 1 + 2 = 5$. If $v = 1$, this constitutes a minimum cost flow.

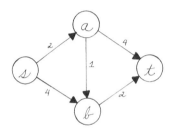

Figure 4.8

Minimum Cost Flow Example

(Arc costs are shown)

Suppose that $v = 2$. By inspection, the only remaining flow augmenting chain is (s,b), (a,b), (a,t) which can accommodate one additional unit from s to t. If this second flow unit is dispatched from s, the first flow unit will be diverted at vertex a so that it takes the path (s,a), (a,t). The second flow unit will replace the first flow unit at vertex b so that it takes the path (s,b), (b,t). The total cost for these two flow units is $a(s,a) + a(a,t) + a(s,b) + a(b,t) = 2 + 4 + 4 + 2 = 12$, which is an increase of 7. This increase of 7 arises from the flow augmenting chain (s,b), (a,b), (a,t) in the following ways:

Cost of +4 for using arc (s,b) in the forward direction

Cost of -1 for using arc (a,b) in the backward direction

Cost of +4 for using arc (a,t) in the forward direction

for a total incremental cost of +7.

Thus, the algorithm must find a flow augmenting chain with the property that the sum of the costs of the forward arcs in the chain less the arc of the costs of the backward arcs in the chain equals p.

4.3 Minimum Cost Flow Algorithm

The algorithm accomplishes this by assigning an integer p(x) to each vertex x in the graph. These vertex numbers p(x) have the properties that $p(s) = 0$, $p(t) = p$, $0 \leq p(x) \leq p$ for all vertices $x \neq s$, $x \neq t$. The algorithm makes flow changes only along arcs (x,y) for which

$$p(y) - p(x) = a(x,y) \qquad (11)$$

If the algorithm finds a flow augmenting chain from s to t consisting entirely of arcs that satisfy equation (11), then it follows that the total incremental cost for each unit dispatched from s to t along this chain equals p.

With this as motivation, we are now prepared to state formally the *minimum cost flow algorithm*.

Minimum Cost Flow Algorithm

Step 1 (Initialization): Initially, let the flow f(x,y) in each arc (x,y) equal zero. Initially, let $p(x) = 0$ for all vertices x.

Step 2 (Deciding Which Arcs Can Have Flow Changes): Let I be the set of all arcs (x,y) for which

$$p(y) - p(x) = a(x,y)$$

and

$$p(x,y) < c(x,y)$$

Let R be the set of all arcs for which

$$p(y) - p(x) = a(x,y)$$

and

$$0 < f(x,y)$$

Let N be the set of all arcs not in I ∪ R. [The arcs in I and R will be the only arcs considered for possible flow changes. Hence, flow changes are only possible on arcs that satisfy equation (11).]

Step 3 (Flow Change): Perform the maximum flow algorithm with I, R, and N as defined above in Step 2. Stop when a total of v flow units have been sent from s to t or when no more flow from s to t is possible for the current composition of sets I, R, and N. If the former occurs first, stop because the terminal flow is a minimum cost flow that sends v units from s to t.

If the latter occurs first, check to see if the current flow is a maximum flow from s to t. (This is done by verifying if the cut generated by the last coloring of the flow augmenting algorithm is saturated.) If so, then stop because no more flow units can be sent from s to t and the terminal flow is a minimum cost flow. If not, then go to Step 4.

Step 4 (Vertex Number Change): Consider the last coloring done by the flow augmenting algorithm. (Recall that the flow augmenting algorithm is a subroutine in the maximum flow algorithm which is used as a subroutine in this algorithm.) Increase by +1 the vertex number $p(x)$ of each uncolored vertex x. (Note that $p(t)$ increases by +1 since t is uncolored since otherwise a flow augmenting chain would have been discovered.) Return to Step 2.

Proof of the minimum cost flow algorithm. We shall prove that the minimum cost flow algorithm does in fact produce a minimum cost flow of v units from s to t by using the complementary slackness conditions of linear programming described in Sec. 1.3.

As mentioned earlier, the minimum cost flow problem can be stated as the linear programming problem given by relations (6) - (10) Let $p(x)$ denote the dual variable associated with conservation of flow equation (7) for vertex x. Let $p(s)$ denote the dual variable for the conservation of flow equation (6) for the source s. Let $p(t)$ denote the dual variable for the conservation of flow equation (8) for the sink t. (Later on, we will show that the dual variables $p(x)$ are the same as the vertex numbers $p(x)$ generated by the algorithm. So, don't worry about this duplicate notation.) Let $\gamma(x,y)$ denote the dual variable for the capacity constraint, relation (9), for arc (x,y). Lastly, let us regard v as a variable rather than a constant.

The dual linear programming problem for the primal linear programming problem in relations (6) - (10) is

4.3 Minimum Cost Flow Algorithm

Minimize

$$\sum_{(x,y)} c(x,y)\, \gamma(x,y) \tag{12}$$

such that

$$-p(s) + p(t) = p \tag{13}$$
$$p(x) - p(y) + \gamma(x,y) \geq -a(x,y) \quad [\text{for all } (x,y)] \tag{14}$$
$$\gamma(x,y) \geq 0 \quad [\text{for all } (x,y)] \tag{15}$$
$$p(x) \text{ unconstrained} \quad (\text{for all } x) \tag{16}$$

Equation (13) is the dual constraint corresponding to the unconstrained primal variable v. Each relation (14) is a dual equation corresponding to a primal variable $f(x,y)$.

The complementary slackness conditions for the primal-dual pair of linear programming problems become

$$p(x) - p(y) + \gamma(x,y) > -a(x,y) \Rightarrow f(x,y) = 0 \tag{17}$$

and

$$\gamma(x,y) > 0 \Rightarrow f(x,y) = c(x,y) \tag{18}$$

If we let

$$p(s) = 0, \; p(t) = p \tag{19}$$

and

$$\gamma(x,y) = \max\{0,\, p(y) - p(x) - a(x,y)\} \tag{20}$$

for all arcs (x,y), then complementary slackness condition (17) becomes

$$p(y) - p(x) < a(x,y) \Rightarrow f(x,y) = 0 \tag{21}$$

since equation (20) implies that $\gamma(x,y) = 0$. Complementary slackness condition (18) becomes

$$p(y) - p(x) > a(x,y) \Rightarrow f(x,y) = c(x,y) \tag{22}$$

Hence from the complementary slackness conditions of linear programming, we need only construct values for $p(x)$ for all vertices x and values for $f(x,y)$ for all arcs (x,y) that satisfy conditions (19), (21), and (22). Of course, the values chosen for $f(x,y)$ must form a feasible solution to the minimum cost flow problem of relations (6) - (10).†

When the algorithm is initialized, $p(x) = 0$ for all x and $p = 0 = p(t)$. Hence, the complementary slcakness conditions are satisfied. We shall now show that the complementary slackness conditions remain satisfied throughout all iterations of the minimum cost flow algorithm. This is accomplished in two parts:

(a) By showing that all flow changes made by the algorithm maintain the complementary slackness conditions

(b) By showing that all vertex number changes maintain the complementary slackness conditions for $p = p(t)$.

[Note that condition (19) is always satisfied by the algorithm and that the values for $f(x,y)$ always yield a feasible flow that satisfies relations (6) - (9).]

The algorithm allows a flow change in arc (x,y) only when $p(y) - p(x) = a(x,y)$. Hence, flow changes cannot disturb the complementary slackness conditions which pertain only to arcs for which $p(y) - p(x) \neq a(x,y)$. Thus, part (a) is verified.

It remains to show that the vertex number changes made by the algorithm do not destroy complementary slackness. (Recall that the algorithm increases a vertex number by +1 only when the vertex could not be colored. If a vertex cannot be colored, then no additional flow units can arrive at that vertex from the source.) If both endpoints of arc (x,y) are colored or if both endpoints of arc (x,y) are uncolored, then $p(y) - p(x)$ remains unchanged and complementary slackness is preserved.

If vertex x is colored and vertex y is uncolored, then we know from the algorithm that one of the following occurs:

(a) $p(y) - p(x) < a(x,y)$
(b) $p(y) - p(x) > a(x,y)$
(c) $p(y) - p(x) = a(x,y)$ and $f(x,y) = c(x,y)$.

---

†Note that the values for $p(x)$, $x \in X$, determine feasible values for all $\gamma(x,y)$ by condition (20).

4.3 Minimum Cost Flow Algorithm

If (a) occurs, then after increasing $p(y)$ by +1, $p(y) - p(x) \leq a(x,y)$ and (21) remains satisfied. If (b) occurs, then after increasing $p(y)$ by +1, $p(y) - p(x)$ remains $> a(x,y)$ and (22) remains satisfied. If (c) occurs, then after increasing $p(y)$ by +1, $p(y) - p(x) > a(x,y)$ and (22) is satisfied.

If vertex x is uncolored and vertex y is colored, then we know from the algorithm that one of the following occurs:

(a) $p(y) - p(x) < a(x,y)$
(b) $p(y) - p(x) > a(x,y)$
(c) $p(y) - p(x) = a(x,y)$ and $ff(x,y) = 0$

If (a) occurs, then after increasing $p(x)$ by +1, $p(y) - p(x)$ remains $< a(x,y)$, and (21) remains satisfied. If (b) occurs, than after increasing $p(x)$ by +1, $p(y) - p(x) \geq a(x,y)$, and (22) remains satisfied. If (c) occurs, then after increasing $p(x)$ by +1, $p(y) - p(x) < a(x,y)$ and (21) is satisfied. This completes the verification of part (b).

Hence, the minimum cost flow algorithm constructs a feasible flow that satisfies all complementary slackness conditions for $p = p(t)$. When the algorithm terminates after dispatching the last flow unit from s to t, the terminal value of $p(t)$ will equal the total incremental cost incurred by the last flow unit. Hence, $p = p(t)$ will be suitably large enough to insure that expression (10) can replace expression (5) as the objective function of the minimum cost flow linear programming problem. Hence, the flow constructed by the algorithm is a minimum cost flow.

The proof is complete except we must show that the minimum cost flow algorithm terminates in a finite number of steps. The minimum cost flow algorithm terminates after the last flow unit has been dispatched from the source to the sink. When this occurs, $p(t)$ equals the total incremental cost incurred by this last flow unit. If all arc costs are finite, positive integers, then at termination $p(t)$ must be a finite positive integer. Hence only $p(t) + 1$ applications of the maximum flow algorithm are required by the minimum cost flow algorithm. As we know from before, the maximum flow algorithm can be modified to terminate in a finite number of steps. Hence, the minimum cost flow algorithm also terminates in a finite number of steps. Q.E.D.

EXAMPLE 3. We shall now apply the minimum cost flow algorithm to find the minimum cost flow for the graph in Fig. 4.9. Note that the first number attached to each arc is the arc cost; the second number attached to each arc is the arc capacity.

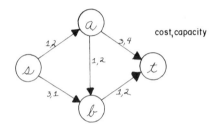

Figure 4.9
Minimum Cost Flow Algorithm

Initially, all vertex numbers are zero and all vertices are uncolored except vertex s, which is always colored. The results of the algorithm are tabulated below:

Iteration	p(s)	p(a)	p(b)	p(t)	Colored arcs	Colored vertices
0	0	0	0	0	none	s
1	0	1	1	1	(s,a)	s,a
2	0	1	2	2	(s,a),(a,b)	s,a,b
3	0	1	2	3	(s,a),(a,b),(b,t)	s,a,b,t

Vertex t has been colored. Send 2 flow units from s to t along path (s,a), (a,b), (b,t). Hence, f(s,a) = 2, f(a,b) = 2, f(b,t) = 2.

3	0	1	2	3	none	s
4	0	2	3	4	(s,b),(a,b)	s,a,b
5	0	2	3	5	(s,b),(a,b),(a,t)	s,a,b,t

4.3 Minimum Cost Flow Algorithm

Vertex t has been colored. Send 1 flow unit from s to t along chain (s,b), (a,b), (a,t). Hence, f(s,a) = 2, f(s,b) = 1, f(a,b) = 1, f(a,t) = 1, f(b,t) = 2.

5	0	2	3	5	none	s

No additional flow units can be sent from s to t since the arcs from the colored vertices to the uncolored vertices [namely, arcs (s,a) and (s,b)] are saturated. Hence, the current flow of three units is a maximum flow with minimum possible cost.

So far, arc costs were assumed to be positive integers. We shall now describe how to modify the minimum cost flow algorithm to accommodate noninteger positive arc costs.

Where did the minimum cost flow algorithm use the assumption that arc costs were integers? Since arc costs were assumed to be positive integers, all flow units sent from s to t incurred a total incremental cost that was a positive integer. Hence, the algorithm needed to examine only integer values of p and vertex numbers were always incremented by the integer amount +1.

When arc costs are not necessarily integer, the total cost of a flow augmenting path from s to t need not be an integer, and hence, the algorithm must examine values for p that are not integers. Which values for p should the algorithm examine? What vertex number increments should the algorithm make?

Suppose the algorithm has been performed for some value of $p = p(t)$. Some arcs will have one colored and one uncolored endpoint. We must consider two cases:

Case 1: If vertex x is colored and vertex y is uncolored, then arc (x,y) is a candidate for coloring only if $(x,y) \in I$ and $p(y)$ can be altered so that $p(y) - p(x) = a(x,y)$. If $(x,y) \in I$, then $p(y) - p(x) \leq a(x,y)$ by condition (22). Hence, $p(y)$ must be increased by $a(x,y) - p(y) + p(x)$ units before arc (x,y) can be colored. Let $\delta(x,y) = a(x,y) - p(y) + p(x)$.

Case 2: If vertex y is colored and vertex x is uncolored, then arc (x,y) is a candidate for coloring only if $(x,y) \in R$ and p(x) can be altered so that p(y) - p(x) = a(x,y). If $(x,y) \in R$, then p(y) - p(x) $\geq$ a(x,y) by condition (21). Hence, p(x) must be increased by p(y) - p(x) - a(x,y) units before arc (x,y) can be colored. Let $\delta(x,y) = p(y) - p(x) - a(x,y)$.

If arc (x,y) does not fit Case 1 or Case 2, let $\delta(x,y) = \infty$.
Let

$$\delta = \min_{(x,y)} \{\delta(x,y)\} > 0 \tag{23}$$

Hence, if the dual number p(x) of each uncolored vertex x is increased by δ then at least one additional arc will be colored, and perhaps a new flow augmenting chain will be discovered. Thus, the value of p(t) will be increased by δ. Similarly, each succeeding vertex number increment can be calculated by determining the smallest increment needed to insure that at least one additional arc can be colored. If $\delta = \infty$, then no additional arcs can be colored. In this case, the current flow is a maximum flow.

Thus, by modifying the vertex number increment procedure, the minimum cost flow algorithm can accommodate noninteger arc costs.

Will the modified algorithm terminate after only a finite number of steps? Yes; since p(t) must always equal the sum of the costs along the arcs of some chain from s to t, and since there are only a finite number of distinct chains from s to t, it follows that p(t) can take only a finite number of different values in the modified algorithm. Hence, the modified algorithm must terminate after only a finite number of steps.

4.4 OUT-OF-KILTER ALGORITHM

The minimum cost flow algorithm has some distinct disadvantages:

1. The algorithm must be initialized with a zero flow and unit by unit work its way to a maximum flow

4.4 Out-of-Kilter Algorithm

2. There is no easy way to salvage the results of this algorithm if we discover after termination that an incorrect arc cost or incorrect arc capacity was used
3. Nonzero lower bounds on the flow in an arc are not permitted, and negative arc costs are not permitted.

In this section, a second algorithm to solve the minimum cost flow problem will be presented. This algorithm, called the *out-of-kilter algorithm*, due to Ford and Fulkerson (1962, p. 162), has none of the above disadvantages of the minimum cost flow algorithm. However, it has certain compensating disadvantages of its own that will be discussed later.

Let $l(x,y)$ be a nonnegative number denoting the smallest number of flow units that must traverse arc (x,y). This quantity is called the *lower capacity* of arc (x,y) to distinguish it from $c(x,y)$ which we shall refer to as the *upper capacity* of arc (x,y).

EXAMPLE 1. The travel agent encountered earlier learns that there must be at least 25 tickets sold before a charter flight from city x to city y can take off. In this situation, the corresponding arc (x,y) must have a lower capacity $l(x,y)$ equal to 25.

EXAMPLE 2. Due to local currency restrictions, a multinational company has various accounts of nonconvertible funds deposited in blocked accounts in various foreign banks. Since these funds can only be spent locally, shipment costs within these countries may be regarded as zero (within limits). Consequently, the costs of the arcs corresponding to shipments through these blocked currency countries may be set equal to zero.

EXAMPLE 3. A Zurich based distributor of medical equipment must route a traveling exhibition of his merchandise through a number of European cities. Also, a number of cities may optionally be placed on the itinerary. The cost of sending the exhibit between any two cities is known and the profit or loss that can be expected from orders coming from each city is also known. What is the best way to route the exhibit to meet all obligations with maximum profit?

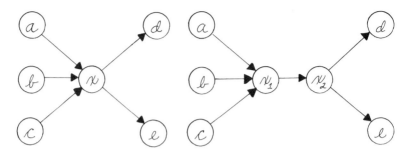

Figure 4.10

Vertex Explosion

[(x_1, x_2) is a city arc. All other arcs are intercity arcs.]

Let each city be represented by a vertex, and let the route between any two cities be represented by an arc. Next, explode each city vertex into two vertices as shown in Fig. 4.10. Let all upper arc capacities equal one; let the lower capacity of all arcs corresponding to cities that must be visited by the exhibition equal 1; let all other lower arc capacities equal 0. Let the cost of a city arc equal the expected profit (positive) or expected loss (negative) from an exhibition in that city. Let the cost of an intercity arc equal the cost of moving the exhibit between the two endpoint cities. A solution to this minimum cost flow problem with nonzero lower arc capacities and negative arc costs may correspond to an optimum exhibition routing. (Chapter 7 describes some complications.) The minimum cost flow for this situation can be found by using the out-of-kilter algorithm described in the following discussion.

The minimum cost flow algorithm solves the minimum cost flow problem when $l(x,y) = 0$ for all arcs (x,y). The out-of-kilter algorithm will solve the minimum cost flow problem for $l(x,y) \geq 0$ for all arcs (x,y). Moreover, the out-of-kilter algorithm allows arc costs to be negative whereas the minimum cost flow algorithm worked only when all arc costs were positive. However, the out-of-kilter algorithm fails if the graph possesses a circuit with infinite arc capacity and negative total cost. In Fig. 4.11, the circuit (a,b), (b,c), (c,a) has infinite capacity and total cost equal to

4.4 Out-of-Kilter Algorithm

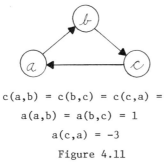

c(a,b) = c(b,c) = c(c,a) =
a(a,b) = a(b,c) = 1
a(c,a) = -3

Figure 4.11

Negative Cost Circuit

1 + 1 - 3 = -1. Consequently, no minimum cost flow exists for this graph since an infinite number of flow units could traverse this circuit each incurring a total cost of -1.

To ease the presentation of the out-of-kilter algorithm, augment the original graph with a return arc (t,s) from the sink to the source. Let all units that were sent from s to t return to the source s via the return arc (t,s). Clearly, any flow in the original graph is equivalent to a flow in the augmented graph, and viceversa. If we are seeking a minimum cost flow that sends v units from s to t, then let $l(t,s) = c(t,s) = v$ and $a(t,s) = 0$. If we are seeking a minimum cost flow that sends the maximum possible number of flow units from s to t, then let $l(t,s) = 0$, $c(t,s) = \infty$, and $a(t,s) = -p$, where (as before p is a large number greater than the cost of sending flow from s to t along the most costly route.

With lower arc capacities and a return arc, the linear programming formulation of the minimum cost flow problem becomes

Minimize

$$\sum_{(x,y)} a(x,y) \, f(x,y) \qquad (24)$$

such that

$$\sum_y f(x,y) - \sum_y f(y,x) = 0 \qquad (25)$$

for all vertices x, and

$$l(x,y) \le f(x,y) \le c(x,y) \qquad (26)$$

for all arcs (x,y).

As before, let the dual variable corresponding to equation (25) for vertex x be denoted by $p(x)$. Let the dual variable corresponding to the lower bound constraint (26) on arc (x,y) be denoted by $\gamma_1(x,y)$. Let the dual variable corresponding to the upper bound constraint (26) on arc (x,y) be denoted by $\gamma_2(x,y)$. With these definitions for the dual variables, the dual linear programming problem is

Maximize
$$\sum_{(x,y)} [\, - c(x,y)\, \gamma_2(x,y) + l(x,y)\, \gamma_1(x,y)] \tag{27}$$

such that
$$- p(x) + p(y) + \gamma_1(x,y) - \gamma_2(x,y) \leq +a(x,y) \tag{28}$$

for all arcs (x,y), and

$$\gamma_1(x,y) \geq 0 \quad [\text{for all } (x,y)] \tag{29}$$

$$\gamma_2(x,y) \geq 0 \quad [\text{for all } (x,y)] \tag{30}$$

The quantity $\gamma_1(x,y) - \gamma_2(x,y)$ appears in each dual constraint (28). Hence, $\gamma_1(x,y) - \gamma_2(x,y)$ can be replaced by $\gamma(x,y)$ which is unconstrained in sign.

Let
$$\gamma(x,y) = a(x,y) + p(x) - p(y) \tag{31}$$

for all arcs (x,y).

Note that equation (31) forces relation (28) to hold with equality for all arcs (x,y).

If $\gamma(x,y) \geq 0$, let

$$\gamma_1(x,y) = \gamma(x,y)$$

and
$$\gamma_2(x,y) = 0.$$

If $\gamma(x,y) < 0$, let

$$\gamma_1(x,y) = 0$$

and
$$\gamma_2(x,y) = -\gamma(x,y).$$

4.4 Out-of-Kilter Algorithm

The complementary slackness conditions of the primal-dual pair of linear programming problems given by (24)-(26) and by (27)-(30) are

$$\gamma_1 > 0 \Rightarrow f(x,y) = l(x,y)$$
$$\gamma_2 > 0 \Rightarrow f(x,y) = c(x,y) \tag{32}$$

Using equation (31), the complementary slackness conditions for the above pair of primal and dual linear programming problems become:

$$p(y) - p(x) < a(x,y) \Rightarrow f(x,y) = l(x,y) \tag{33}$$

and

$$p(y) - p(x) > a(x,y) \Rightarrow f(x,y) = c(x,y) \tag{34}$$

Note the similarity between complementary slackness conditions (21) and (22) and the above complementary slackness conditions.

For convenience later on, let us define

$$\alpha(x,y) = a(x,y) - p(y) + p(x) \tag{35}$$

for all arcs (x,y). Using (35), conditions (33) and (34) become:

$$\alpha(x,y) > 0 \Rightarrow f(x,y) = l(x,y) \tag{36}$$

and

$$\alpha(x,y) < 0 \Rightarrow f(x,y) = c(x,y) \tag{37}$$

To solve the general minimum cost flow problem described above, we need only construct a flow that satisfies equation (25) and vertex numbers that satisfy conditions (36) and (37). (Note that relations (28)-(30) are always satisfied.)

Suppose that we select any flow $f(x,y)$ that satisfies equations (25). Thus, the net flow into each vertex is zero. Furthermore, suppose we arbitrarily select any set of vertex numbers $p(x)$. For the flow values and vertex numbers specified above, each arc (x,y) can fall into one of nine different situations:

Situation				Kilter Number
I.	$\alpha(x,y) < 0$ and	$f(x,y) < c(x,y)$		$\alpha(x,y)[f(x,y) - c(x,y)]$
II.	$\alpha(x,y) < 0$ and	$f(x,y) = c(x,y)$		0
III.	$\alpha(x,y) < 0$ and	$f(x,y) > c(x,y)$		$f(x,y) - c(x,y)$

Situation				Kilter Number
IV.	$\alpha(x,y) = 0$	and	$f(x,y) < l(x,y)$	$l(x,y) - f(x,y)$
V.	$\alpha(x,y) = 0$	and	$l(x,y) \leq f(x,y) \leq c(x,y)$	0
VI.	$\alpha(x,y) = 0$	and	$f(x,y) > c(x,y)$	$f(x,y) - c(x,y)$
VII.	$\alpha(x,y) > 0$	and	$f(x,y) < l(x,y)$	$l(x,y) - f(x,y)$
VIII.	$\alpha(x,y) > 0$	and	$f(x,y) = l(x,y)$	0
IX.	$\alpha(x,y) > 0$	and	$f(x,y) > l(x,y)$	$\alpha(x,y)[f(x,y) - l(x,y)]$

Next to each of the nine different situations is listed the *kilter number* of the situation. The kilter number denotes the amount by which an arc in that situation is said to be *out-of-kilter*. Denote the kilter number of arc (x,y) by k(x,y). Inspection of the above nine situations will show that $k(x,y) \geq 0$. Let K denote the sum of the kilter numbers of all the arcs in the graph. Note that the kilter number of each arc that satisfies the complementary slackness conditions is zero. Moreover, if an arc does not satisfy the complementary slackness conditions then its kilter number is positive.

The basic idea underlying the out-of-kilter algorithm is to successively reduce to zero the kilter number of each arc without increasing the kilter number of any other arc. When this has been accomplished, then every arc will satisfy the complementary slackness conditions and the current flow will be a maximum flow with minimum cost.

How does the out-of-kilter algorithm reduce the kilter number of an arc to zero? For example, suppose that arc (x,y) is out-of-kilter, that is, k(x,y) > 0. The algorithm determines if an increase or a decrease in the flow in arc (x,y) needed to put arc (x,y) into kilter. If a flow increase is needed, then the algorithm searches for a chain from y to x along which flow can be sent without increasing the kilter number of any arc in the graph. If such a chain is discovered by the algorithm, then this chain along with arc (x,y) forms a cycle. Flow can be sent around this cycle thereby increasing the flow in arc (x,y) and not increasing any arc kilter number. If such a flow change still does not put arc (x,y) into kilter, then

4.4 Out-of-Kilter Algorithm

this process is repeated until arc (x,y) is in kilter or until no more such chains can be discovered by the algorithm. When the latter occurs, the algorithm increases some of the vertex numbers (like in the minimum cost flow algorithm) and again looks for a flow accepting chain from y to x. The vertex numbers are increased so that no arc kilter number increases. Ultimately, arc (x,y) either goes into kilter or the algorithm shows that no flow exists that simultaneously satisfies all arc upper and lower capacity requirements.

On the other hand, if a flow decrease is needed to put arc (x,y) into kilter, then the algorithm repeats the same procedure as described above except that now the algorithm searches for a chain from x to y. This process is repeated until every arc is in kilter.

With this as motivation, we can now present the *out-of-kilter algorithm*.

Out-of-Kilter Algorithm

Step 1 (Initialization): Select any set of flow values f(x,y) such that the net flow into each vertex in the graph is zero, i.e., so that equation (25) is satisfied. This flow need not satisfy the upper and lower capacity requirement, relation (26), on each arc. Also, select any set of values for the vertex numbers p(x).

Step 2 (Determining Kilter Numbers): For each arc (x,y) in the graph, calculate $\alpha(x,y)$ and k(x,y) as defined by equations (35) and (38). Stop if all k(x,y) = 0.

Step 3 (Arc Classification): Classify each arc (x,y) as increasable or decreasable as follows:

Arc (x,y) is decreasable if

(a) $\alpha(x,y) \geq 0$ and f(x,y) > l(x,y)

or

(b) $\alpha(x,y) \leq 0$ and f(x,y) > c(x,y).

Arc (x,y) is increasable if

(a) $\alpha(x,y) \geq 0$ and f(x,y) < l(x,y)

or

(b) $\alpha(x,y) \leq 0$ and $f(x,y) < c(x,y)$.

Let R denote the set of reducible arcs; let I denote the set of increasable arcs.

If arc $(x,y) \in R$, then let

$r(x,y) = f(x,y) - l(x,y)$ if $\alpha(x,y) \geq 0$
$r(x,y) = f(x,y) - c(x,y)$ if $\alpha(x,y) < 0$

If arc $(x,y) \in I$, then let

$i(x,y) = l(x,y) - f(x,y)$ if $\alpha(x,y) > 0$
$i(x,y) = c(x,y) - f(x,y)$ if $\alpha(x,y) \leq 0$.

Select any arc (x,y) for which $k(x,y) > 0$. If $(x,y) \in I$, then let vertex y be called s, and let vertex x be called t. If $(x,y) \in R$, then let vertex x be called s, and let vertex y be called t. (Note that since $k(x,y) > 0$, arc (x,y) cannot belong to both I and R.)

Step 4 (Maximum Flow Subroutine): Using the set I and R and values for $i(x,y)$ and $r(x,y)$ defined above, perform the maximum flow algorithm to send flow units from s to t. Balance this flow by making a compensating flow change in the out-of-kilter arc joining s and t. Do this, until enough flow units have been sent from s to t so that the arc joining s and t is in kilter or until no more flow units can be sent from s to t.

If the former occurs first, then arc (x,y) has been placed into kilter. Return to Step 2. If the latter occurs first, then go to Step 5.

Step 5 (Vertex Number Increases): The maximum flow algorithm subroutine of Step 4 stopped without being able to find a flow augmenting chain from s to t. Let C denote the set of vertices colored during the last iteration of the flow augmenting algorithm subroutine of the maximum flow algorithm. Clearly, $s \in C$. Let $\bar{C}$ denote the set of uncolored vertices. Clearly, $t \in \bar{C}$. Define two sets of arcs:

$$A_1 = \{(x,y): x \in C, y \in \bar{C}, \alpha(x,y) > 0, f(x,y) \leq c(x,y)\}$$
$$A_2 = \{(y,x): x \in C, y \in \bar{C}, \alpha(y,x) < 0, f(y,x) \geq l(y,x)\}$$
(39)

4.4 Out-of-Kilter Algorithm

If A_1 is empty, let $\delta_1 = \infty$. Otherwise, let

$$\delta_1 = \min_{A_1}\{\alpha(x,y)\} > 0. \tag{40}$$

If A_2 is empty, let $\delta_2 = \infty$. Otherwise, let

$$\delta_2 = \min_{A_2}\{\alpha(x,y)\} > 0 \tag{41}$$

Let

$$\delta = \min\{\delta_1, \delta_2\} > 0. \tag{43}$$

If $\delta = \infty$, stop because no feasible flow exists for this graph. If $\delta < \infty$, then replace $p(x)$ by $p(x) + \delta$ for all $x \in \bar{C}$. Return to Step 3.

The terminal values of $f(x,y)$ represent a minimum cost flow.

Proof of the out-of-kilter algorithm: To prove the out-of-kilter algorithm, we must show that

(1) The algorithm terminates with a flow that satisfies (25).
(2) That is a minimum cost flow.
(3) In a finite number of steps.
(4) Or that no feasible flow exists for this graph.

Proof of (1): The algorithm terminates with a flow since it is initialized with a flow that satisfies (25) and maintains a flow that satisfies (25) during all flow changes performed by the algorithm.

Proof of (2): There are only two ways that the algorithm can stop: all arcs are in kilter or $\delta = \infty$ at some iteration. If all arcs are in kilter, then the terminal flow values and vertex number values satisfy the complementary slackness conditions, (36) and (37) and the terminal flow is a minimum cost flow.

Proof of (4): If the algorithm terminates because $\delta = \infty$ at some iteration of Step 5, then it is claimed that it is impossible for any flow to satisfy simultaneously all upper and lower arc capacities.

Consider the cut set of arcs with one endpoint in C and the other endpoing in $\bar{C}$. No arc from C to $\bar{C}$ is increasable since otherwise its endpoint in $\bar{C}$ could be colored. Likewise, no arc from $\bar{C}$ to C is decreasable since otherwise its endpoint in $\bar{C}$ could be colored. Moreover, since $\delta = \infty$, both sets A_1 and A_2 as defined in equation (39) are empty sets.

Since $\delta = \infty$, it follows that any arc (x,y) from C to $\bar{C}$ must carry a flow $f(x,y) \geq c(x,y)$. Also, any arc (x,y) from $\bar{C}$ to C must carry a flow $f(x,y) \leq l(x,y)$. Furthermore, for the out-of-kilter arc joining s and t this inequality is a strict inequality.

Add together the equation (25) for each vertex in C. The sum of these equations can be interpreted as follows: the net flow out of C must equal zero. The net flow out of C consists of the total flow from the vertices in C to the vertices in $\bar{C}$, denoted by $f(C,\bar{C})$, less the total flow from vertices in $\bar{C}$ to the vertices in C, denoted by $f(\bar{C},C)$. Thus,

$$f(C,\bar{C}) - f(\bar{C},C) = 0 \qquad (44)$$

As noted above, $f(\bar{C},C) \leq l(\bar{C},C)$ and $f(C,\bar{C}) \geq c(C,\bar{C})$ and at least one of these inequalities is a strict inequality, where $l(X,Y)$ denotes the sum of the lower capacities of the arcs from set X to set Y and $c(X,Y)$ denotes the sum of the upper capacities of the arcs from set X to set Y. Hence,

$$c(C,\bar{C}) - l(\bar{C},C) < 0$$

and

$$c(C,\bar{C}) < l(\bar{C},C),$$

which implies that the smallest flow $l(\bar{C},C)$ that must flow into C exceeds the maximum flow $c(C,\bar{C})$ that can flow out of C. Hence, no feasible flow exists for $\delta = \infty$.

Proof of (3): It remains to show that the algorithm terminates after a finite number of steps.

First, observe that whenever the algorithm makes a change in a flow value or vertex number that no arc kilter number is increased. Hence, once an arc is in kilter, this arc remains in kilter. Thus,

4.4 Out-of-Kilter Algorithm

if the algorithm required an infinite number of steps, then there would be some arc that would require an infinite number of steps to be placed into kilter.

To place a given arc into kilter, the algorithm performs a sequence of flow augmentations and vertex number increases. If all arc capacities were assumed to be integers, only a finite number of flow augmentations can occur since each augmentation changes the flow in the out-of-kilter arc by at least one unit. Hence, if an infinite number of steps were required to put an arc into kilter, then an infinite number of consecutive vertex number increases must occur between two flow augmentations. (If not all arc capacities are integers, the finite termination modification of the maximum flow algorithm must be employed. In this case, the number of flow augmentations is again finite by reasoning similar to the reasoning used in the proof of the finite termination modification. Namely, between any two flow augmentations in which arc (x,y) is a bottleneck arc, the minimum number of arcs in a flow augmenting chain from s to t must increase by at least two in the subgraph consisting of only the arcs in I $\cup$ R and cannot decrease in any other subgraph of the original graph. Since only a finite number of subgraphs can be derived from the original graph, only a finite number of flow augmentations are possible.)

Observe that a vertex number increase results either in the coloring of at least one additional vertex or the deletion of at least one arc from the set $A_1 \cup A_2$. Hence, ultimately since there are only a finite number of vertices, it follows that ultimately vertex t must be colored or sets A_1 and A_2 become empty resulting in $\delta = \infty$. Thus, the algorithm cannot encounter an infinite number of consecutive vertex number increases. Hence, the algorithm must terminate in a finite number of steps. Q.E.D.

In the beginning of this section, we mentioned that the out-of-kilter algorithm did not have certain disadvantages that the minimum cost flow algorithm has, but that the out-of-kilter algorithm had certain disadvantages of its own. What is the main disadvantage of the out-of-kilter algorithm? It is that one must initialize the algorithm with a set of vertex number values. In many cases, one has no idea of initial vertex number values. This could result in arbitrary choices of the initial vertex number values that would result in many arcs being heavily out-of-kilter and consequently requiring many flow augmentations to get into kilter.

On the other hand, the out-of-kilter algorithm does allow negative arc costs and lower arc capacities. Moreover, if an arc cost or capacity changes, the previous optimal solution could be used to initialize the out-of-kilter algorithm to solve the minimum cost flow problem with the updated values.

4.5 DYNAMIC FLOW ALGORITHMS

In the preceding four sections, we studied flows that obeyed certain requirements dictated by the arc capacities and the arc costs. In this section, we shall consider yet another arc requirement, namely arc traverse time, and we shall study flows in which all the flow units must make the trip from the source to the sink within a given amount of time.

Associate each arc (x,y) in graph $G = (X, A)$ a positive integer $a(x,y)$ that denotes the number of time periods required by a flow unit to travel across arc (x,y) from x to y. The quantity $a(x,y)$ is called the *traverse time* of arc (x,y). (Yes, previously $a(x,y)$ was used to denote the cost of arc (x,y). As will be seen later, both arc cost and traverse time will fulfill the same role in algorithms, and hence we have chose to identify them with the same symbol. This will become clearer later.) Let $c(x,y,T)$ denote the maximum number of flow units that can enter arc (x,y) at the start of time period T, for $T = 0, 1, \ldots$. A *dynamic flow* from s to t in graph G is any flow from s to t in graph G that obeys all arc capacity requirements at all times. Thus, a dynamic flow from s to t is any flow from s to t in which not more than $c(x,y,T)$ flow units enter arc (x,y) at the start of time period T, for all arcs (x,y) and all T. Note that in a dynamic flow, units may be departing from the source at time $0, 1, 2, \ldots$.

4.5 Dynamic Flow Algorithms

A *maximum dynamic flow for p time periods* from s to t is any dynamic flow from s to t in which the maximum possible number of flow units arrive at the sink t during the first p time periods.

EXAMPLE 1. Our friend, the travel agent, must send 75 passengers from Springfield to Istanbul within the next 48 hours. This problem can be rephrased as a maximum dynamic flow problem as follows: Let Springfield represent the source, and let Istanbul represent the sink. Let each airport between Springfield and Istanbul be represented by a vertex. Join vertices x and y by an arc (x,y) if there is a nonstop flight from airport x to airport y. Let the traverse time of this arc be equal to the flight time between these two airports rounded up to the next hour. (Airport transfer time should also be included in flight time.) Let the capcaity c(x,y,T) for arc (x,y) at time T equal the number of seats available on the flight from airport x to airport y that starts at time T. If no such flight exists, then let c(x,y,T) = 0.

The travel agent's problem is solved if there exists a flow of 75 units from the source to the sink of this graph within 48 time periods, and if he can construct such a flow.

Obviously, the problem of finding a maximum dynamic flow is more complex than the problem of finding a maximum flow since the dynamic flow problem requires that we keep track of when each unit travels through an arc so that no arc's entrance capacity is violated at any time. Happily, this additional complication can be resolved by rephrasing the dynamic flow problem into a static (nondynamic) flow problem on a new graph called the time-expanded replica of the original graph.

The *time-expanded replica of graph* $G = (X, A)$ *for p time periods* is a graph G_p whose vertex set is

$$X_p = \{x_i : x \in X, i = 0, ., \ldots, p\} \quad (45)$$

and whose arc set is

$$A_p = \left\{ (x_i, y_j) : (x,y) \ A, \ i = 0, 1, \ldots, p - a(x,y), \atop j = i + a(x,y) \right\} \quad (46)$$

Let

$$c(x_i, y_j) = c(x,y,i).$$

Note that the vertex set X_p of graph G_p consists of duplicating each vertex in X once for each time period. An arc joins vertices x_i and y_j only if it takes $j - i$ time periods to travel from vertex x to vertex y. Thus, a flow unit leaving vertex x along arc (x,y) in graph G at time 5 and taking 8 time periods to arrive at vertex y is represented in graph G_p as a unit flowing along the arc (x_5, y_{13}). Figure 4.12 shows the time-expanded replica of a graph for 6 time periods.

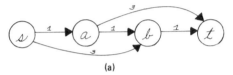

(a)

(a) Original Graph

(All arc capacities equal one. Traverse times
are given next to each arc.)

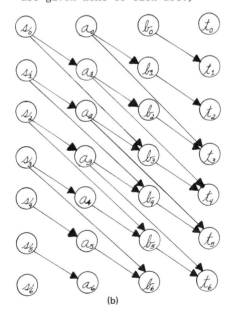

(b)

Figure 4.12

(b) Time-Expanded Replica Graph (All arc capacities are one.)

p = 6

4.5 Dynamic Flow Algorithms

Clearly, any dynamic flow from s to t in graph G is equivalent to a flow from the sources to the sinks in G_p, and viceversa. Figure 4.13 shows a dynamic flow and its static equivalent for the graph in Fig. 4.12.

	DYNAMIC FLOW			STATIC FLOW	
Path	Departure Time	Amount		Path	Amount
s,a,b,t	0	1 unit	$\longleftrightarrow$	s_0, a_1, b_2, t_3	1 unit
s,a,b,t	1	1 unit	$\longleftrightarrow$	s_1, a_2, b_3, t_4	1 unit
s,a,t	2	1 unit	$\longleftrightarrow$	s_2, a_3, t_6	1 unit

Figure 4.13

Equivalence Between Dynamic and Static Flows

Since each dynamic flow is equivalent to a static flow in the time-expanded replica graph, a maximum dynamic flow for p time periods can be found simply by finding a maximum flow in the time-expanded replica for p time periods using the maximum flow algorithm. Thus, no additional algorithms are required to solve the maximum dynamic flow problem. However, if p is very large, then the graph G_p becomes very large and the number of calculations required to find a maximum flow of graph G_p becomes prohibitively large.

Happily, Ford and Fulkerson (1962, p. 142) have devised an algorithm, called the *maximum dynamic flow algorithm*, that generates a maximum dynamic flow much more efficiently than the method suggested above. However, the maximum dynamic flow algorithm works only when the entrance capacity of each arc remains unchanged through time, i.e., when

$c(x,y,T) = c(x,y)$ (for all $T = 0, 1, \ldots, p$)

for all arcs (x,y).

The maximum dynamic flow algorithm uses the minimum cost flow algorithm as a subroutine. Recall that the minimum cost flow algorithm first sends as many units as possible from s to t that have total incremental cost of 1 each. Next, the minimum cost flow algorithm sends as many units as possible from s to t that have total incremental cost of 2 each, etc. until a maximum flow has been attained. Let $F_1, F_2, \ldots, F_p$ denote the resulting sequence of flows generated by the minimum cost flow algorithm. (The details of how F_i is generated from F_{i-1} were presented in Section 3.) Each flow F_i consists of a set of paths $f_{i,1}, f_{i,2}, \ldots, f_{i,r_i}$ from s to t taken by the units sent from s to t by flow F_i. Let $n_{i,j}$ denote the number of flow units that would follow path $f_{i,j}$ in flow F_i. Clearly, no flow path $f_{i,j}$ from s to t has a total cost greater than i, since otherwise, the incremental cost of using this path would be greater than i and this path could not have been generated by the minimum cost flow algorithm. Let $a(f_{i,j})$ denote the total cost of flow path $f_{i,j}$.

With these definitions in mind, we can now state the *maximum dynamic flow algorithm*.

Maximum Dynamic Flow Algorithm

This algorithm generates a maximum dynamic flow for p time units from the source s to the sink t in a graph $G = (X, A)$ for which $c(x,y,T) = c(x,y)$ for all $T = 0, 1, \ldots, p$ and for all $(x,y) \in \Gamma$.

Step 1: Let the cost of using arc (x,y) equal the transit time $a(x,y)$ for all arcs (x,y). Perform the minimum cost flow algorithm on graph G stopping after the flow F_p has been generated.

Flow F_p consists of flow paths $f_{p,1}, f_{p,2}, \ldots, f_{p,r_p}$ from s to t respectively carrying $n_{p,1}, n_{p,2}, \ldots, n_{p,r_p}$ flow units from s to t. The decomposition of F_p into flow paths is a by-product of the minimum cost flow algorithm.

Step 2: For $j = 1, 2, \ldots, r_p$, send $n_{p,j}$ flow units from s to t along flow path $f_{p,j}$ starting out from s at time periods $0, 1, \ldots, p - a(f_{p,j})$. The resulting flow is a maximum dynamic flow for p time periods.

Thus, the maximum dynamic flow algorithm consists of performing the minimum cost flow algorithm using the arc traverse times as the arc costs. The terminal flow generated by the minimum cost flow

4.5 Dynamic Flow Algorithms

algorithm is decomposed into paths from s to t. Lastly, flow is sent along each of these paths starting at time period 0, 1, ..., until flow can no longer reach the sink by time period p. Figure 4.14 shows a maximum dynamic flow for p = 4, 5, 6 for the graph in Fig. 4.12.

The flow generated by the maximum dynamic flow algorithm is called a *temporally repeated* flow for the obvious reason that it consists of repeating shipments along the same flow paths from s to t. Of course, the maximum dynamic flow for p time periods generated by the maximum dynamic flow algorithm need not be the only possible maximum dynamic flow for p time periods. However, as shown by the algorithm, there is always one maximum dynamic flow for p time periods that is a temporally repeated flow.

Proof: To prove the maximum dynamic flow algorithm, we must show

(a) That the algorithm constructs a flow
(b) That this flow is a maximum dynamic flow for p time periods
(c) That the algorithm terminates in a finite number of steps.

Proof of (a): The algorithm sends flow units from s to t along the paths in F_p. Except for the source and the sink, the flow into a vertex equals the flow out of that vertex. Moreover, no flow units are dispatched unless they can reach the sink before time period p.

How many flow units enter arc (x,y) during any given time period? Flow F_p cannot send more than c(x,y) flow units across arc (x,y), hence the flow paths $f_{p,1}$, $f_{p,2}$, ..., f_{p,r_p} cannot send more than a total of c(x,y) flow units across arc (x,y). Hence, not more than c(x,y) flow units can enter arc (x,y) during any given time period. Hence, the dynamic flow constructed by the maximum dynamic flow algorithm obeys all arc capacities.

Hence, since all flow units start at s and arrive at t before time period p, violate no arc capacities and do not holdover at any intermediate vertex, the algorithm produces a dynamic flow.

Results of the Minimum Cost Flow Algorithm

p	F_p
0	0
1	0
2	0
3	1 unit along path s, a, b, t
4	1 unit along path s, a, b, t
5	1 unit along path s, b, t
	1 unit along path s, a, b
6	1 unit along path s, b, t
	1 unit along path s, a, t

Maximum Dynamic Flow for p Time Periods

$p = 0$ No flow.

$p = 1$ No flow.

$p = 2$ No flow.

$p = 3$ Send one unit along path s, a, b, t starting at time 0 to arrive at the sink at time 3.
Total flow = 1 unit.

$p = 4$ Send one unit along path s, a, b, t starting at time 0 to arrive at the sink at time 3.

Send one unit along path s, a, b, t starting at time 1 to arrive at the sink at time 4.
Total flow = 2 units.

$p = 5$ Send one unit along path s, b, t starting at time 0 to arrive at the sink at time 4.
Send one unit along path s, b, t starting at time 1 to arrive at the sink at time 5.
Send one unit along path s, a, t starting at time 0 to arrive at the sink at time 4.
Send one unit along path s, a, t starting at time 1 to arrive at the sink at time 5.
Total flow = 4 units.

$p = 6$ Send one unit along path s, b, t starting at times 0, 1, 2 to arrive at the sink at times 4, 5, 6.
Send one unit along path s, a, t starting at times 0, 1, 2 to arrive at the sink at times 4, 5, 6.
Total flow = 6 units.

Figure 4.14

Example of a Maximum Dynamic Flow Algorithm

4.5 Dynamic Flow Algorithms

Proof of (b): To show that the flow generated by the algorithm is a maximum dynamic flow, we shall consider its equivalent in the time-expanded replica graph G_p. We shall accomplish the proof by showing that this static flow saturates a cut of arcs separating the sources from the sinks in graph G_p.

Let $p(x)$ denote the value of the dual variable for vertex x just before the $(p + 1)$-th iteration of the minimum cost flow algorithm begins. Thus, $p(t) = p + 1$. Let

$$C = \{x_i: x_i \in X_p, p(x) \leq i\}$$

Note that since $p(s) = 0$, all source vertices $s_0, s_1, \ldots, s_p$ are members of set C. Also, since $p(t) = p + 1$, no sink vertex $t_0, t_1, \ldots, t_p$ is a member of set C. The set of all arcs with one endpoint in C and the other endpoint not in C form a cut K that separates the sources from the sinks.

Suppose that arc $(x_i, y_j) \in K$, $x_i \in C$, $y_j \notin C$, then $a(x,y) = j - i < p(y) - p(x)$ from the definition of cut K. It follows from the complementary slackness conditions of the minimum cost flow algorithm that $f(x,y) = c(x,y)$. Hence, the flow paths that comprise F_p must saturate arc (x,y). Each of these flow paths that uses arc (x,y) has a total traverse time (equivalently total cost) not greater than p time periods. Moreover, each of these paths is short enough so that flow units traveling along them can reach vertex x at time i and still arrive at the sink before or during the p-th time period. Thus, arc (x_i, y_j) is saturated in graph G_p.

By a similar argument, if $x_i \notin C$ and $y_j \in C$, then arc (x_i, y_j) is empty.

Thus, each cut arc in graph G_p from the source side of the cut to the sink side of the cut is saturated and each arc from the sink side of the cut to the source side of the cut is empty. Hence, this cut K is saturated and the flow produced by the algorithm corresponds to a maximum flow in graph G_p. Thus, the dynamic flow produced by the algorithm is a maximum dynamic flow for p time periods.

Proof of (c): The minimum cost flow algorithm terminates in a finite number of steps. Hence, the maximum dynamic flow algorithm must terminate in a finite number of steps since it consists of one performance of the minimum cost flow algorithm and a temporal repetition of flow along a finite number of flow paths from s to t. Q.E.D.

The preceding discussion of dynamic flows did not consider the possibility of a flow unit stopping to rest or holding over at a vertex for one or more time periods before continuing its journey to the sink. For example, some of the passengers from Springfield to Istanbul might wish to break their journey for a few days at an intermediate city.

If holdovers are permitted, then graph G_p must be revised to contain arcs of the form (x_i, x_{i+1}) so that flow units holding over at vertex x can depart from vertex x at a later time period.

If holdovers are permitted, how does the maximum dynamic flow for p time periods change? Obviously, the possibility of having holdovers cannot decrease the maximum number of flow units that can be sent from s to t in p time periods. In fact, it is easy to show that the maximum dynamic flow for p time periods remains unchanged: Suppose holdovers are permitted and graph G_p is augmented with additional arcs as described above. Consider the flow generated by the maximum dynamic flow algorithm and the cut K that this flow saturates in the original unaugmented graph G_p. This cut K is also a cut of the augmented graph G_p and is still saturated by the maximum dynamic flow produced by the maximum dynamic flow algorithm, since each holdover arc in K is directed from $\bar{c}$ to c and carries no flow. Thus, the possibility of holdovers does not change the maximum dynamic flow.

Lexicographic Dynamic Flows

Let V(p) denote the maximum number of flow units that can be sent from the source to the sink in graph G within p time periods. Obviously,

4.5 Dynamic Flow Algorithms

$$V(1) \le V(x) \le \ldots \le V(p) \le V(p+1)$$

For the special case when arc transit times remained stationary (i.e., were the same during all time periods), the maximum dynamic flow algorithm constructed a maximum dynamic flow for p time periods by temporally repeating the flow F_p obtained after the p-th iteration of the minimum cost flow algorithm. Thus,

$$V(p) = \sum_{i=1}^{i=r_p} [p - a(f_{p,i})] n_{p,i} \tag{47}$$

since each flow path $f_{p,i}$ in F_p is repeated $p - a(f_{p,i})$ times and carries $n_{p,i}$ flow units. Let V_p denote the number of flow units sent from s to t at the end of the p-th iteration of the minimum cost flow algorithm. Then, rewriting equation (47) yields

$$V(p) = pV_p - \sum_{i=1}^{i=r_p} a(f_{p,i}) n_{p,i} = pV_p - \sum_{(x,y)} a(x,y) f_p(x,y) \tag{48}$$

Thus, we can conclude that

$$v(p+1) - v(p) = (p+1)V_{p+1} - pV_p - \sum_{(x,y)} a(x,y)[f_{p+1}(x,y) - f_p(x,y)]$$

$$= V_{p+1} \ge V_p \tag{49}$$

Note that $V(p+1) - V(p) = V_p$ if, and only if, flow F_{p+1} is identical to flow F_p. Consequently, *as p increases the maximum dynamic flow increased by at least V_p units.*

An *earliest arrival flow for p time periods* is any maximum dynamic flow for p time periods in which V(i) flow units arrive at the sink during the first i time periods for i = 0, 1, ..., p. Thus, an earliest arrival flow (if it exists) if a flow that is simultaneously a maximum dynamic flow for 0, 1, ..., p time periods. It is logical to call such a flow an earliest arrival flow since it would be impossible for any flow units to arrive any earlier at the sink.

EXAMPLE 2. The travel agent of the previous example wishes as many passengers from Springfield to arrive in Istanbul within the first hour, within the first two hours, ..., within the first 47 hours,

within the first 48 hours. Rephrased in graph terms, the travel
agent is seeking an earliest arrival flow from Springfield to Istanbul for 48 time periods.

THEOREM 4.1. An earliest arrival flow for p time periods always exists for graph G.

Proof: Use the maximum flow algorithm to construct a maximum flow from the sources to the sink in graph G_p for zero time periods. Call this flow MDF_0 since it is a maximum dynamic flow for zero time periods.

Next, starting with MDF_0 in graph G_p generate a maximum flow from the sources into sinks t_0 and t_1. Clearly, this flow is a maximum dynamic flow for one time period. Call this flow MDF_1. Note that when the maximum flow algorithm is performed to generate MDF_1 from MDF_0 that no flow unit entering sink t_0 is ever rerouted to sink t_1 since the algorithm will never reroute a flow unit already at a sink. However, the route taken by a unit arriving at sink t_0 might be altered when MDF_1 is constructed. Consequently, MDF_1 will be a maximum dynamic flow for zero time periods as well as for one time period.

In a similar way, MDF_2 can be generated from MDF_1 and MDF_2 will be a maximum dynamic flow for zero, one and two time periods.

Repeat this process to generate MDF_p which is an earliest arrival flow. Q.E.D.

The above proof not only demonstrates that an earliest arrival flow always exists for any graph G and for any p = 0, 1, ..., but also shows how to construct an earliest arrival flow. For large graphs or for large values of p, this construction procedure requires an excessive and perhaps prohibitive number of calculations. Fortunately, there is an efficient algorithm to construct an earliest arrival flow for the special case when arc capacities are stationary through time, i.e., when $c(x,y,T) = c(x,y)$ for all T and all arcs (x,y)

4.5 Dynamic Flow Algorithms

This algorithm is called the earliest arrival flow algorithm (Minieka, 1973; Wilkinson, 1971).

Before discussing the earliest arrival flow algorithm, let us explore conditions under which the maximum dynamic flow algorithm will and will not construct a maximum dynamic flow that is also an earliest arrival flow.

Consider the graph in Fig. 4.12 and the maximum dynamic flow for p = 6 for this graph given in Fig. 4.14. This flow consists of 6 units. One flow unit is dispatched at time 0, 1, 2 along each of the two flow paths (s,a,t) and (s,b,t). Thus, two flow units arrive at the sink at time 4,5,6. This flow cannot be an earliest arrival flow because no flow units arrive at the sink at time 3. Note that during the fifth iteration (p = 5) of the minimum cost flow algorithm performed on this graph the flow augmenting chain (s,b,a,t) was discovered. This chain has total cost (traverse time) equal to 3 - 1 + 3 = 5. This augmenting chain transformed flow path (s,a,b,t) of total cost 1 + 1 + 1 = 3 into path (s,a,t) with total cost 1 + 3 = 4. Hence, the units traveling this path will now be detoured to a path that costs one unit more.

The key idea behind the earliest arrival flow algorithm is to channel as many flow units as possible along shorter paths, like (s,a,b,t), before these paths are superseded by longer paths, like (s,a,t).

To facilitate the presentation of the earliest arrival algorithm, we shall assume that all c(x,y) are integers, and replace each arc (x,y) by c(x,y) replicas of itself each with a capacity of one unit. Thus, since all arc capacities are one, all flow paths will have a capacity of one, and each arc will be either saturated or empty.

As before, let F_i denote the flow produced after the i-th iteration of the minimum cost flow algorithm. Let $f_{i,1}$, $f_{i,2}$, ..., f_{i,r_i} denote the flow paths that constitute flow F_i. By the above assumption, each of these flow paths carries one flow unit. (Thus, $n_{i,j} = 1$ for all i and all j.) Let $P_{i,j}$ denote the j-th flow augmenting chain from s to t discovered during the i-th iteration of the minimum cost flow algorithm. In the example in Fig. 4.14, $P_{3,1}$ = (s,a,b,t), and $P_{5,1}$ = (s,b,a,t).

With these definitions and motivation, we are now prepared to state the earliest arrival flow algorithm.

Earliest Arrival Flow Algorithm

Step 1 (Initialization): Initially, there is no flow in any arc. Perform the minimum cost flow algorithm for p iterations. Record each flow F_i, each set of flow paths, $f_{i,1}$, $f_{i,2}$, ..., f_{i,r_i}, and each breakthrough path $P_{i,j}$ for i = 0, 1, ..., p and j = 1, 2, ..., w_i.

Step 2 (Flow Construction): Repeat the following procedure for δ = 0, 1, ..., p:

Consider the sequence

$$P_{1,1}, P_{1,2}, \ldots, P_{1,w_1}, \ldots, P_{\delta,1}, \ldots, P_{\delta,r_\delta}$$

At time period δ - i dispatch one flow unit from s to t along each chain $P_{i,j}$ in the above sequence. Let the flow unit spend a(x,y) time periods in each forward arc (x,y) in chain $P_{i,j}$ and -a(x,y) time periods in each backward arc in chain $P_{i,j}$. Label each forward arc in chain $P_{i,j}$ with the time that this flow unit enters the arc, and remove for each backward arc the label (if it exists) that denotes the time that the flow unit would leave the backward arc.

Upon termination of the above procedure, the labels represent the times that a flow unit enters an arc, and correspond to an earliest arrival flow for p time periods.

EXAMPLE 3. We shall now use the earliest arrival flow algorithm to construct an earliest arrival flow for the graph in Fig. 4.12 for p = 6. Recall that when the minimum cost flow algorithm was performed on this graph only two flow augmenting chains were discovered, namely $P_{3,1}$ = (s,a,b,t) and $P_{5,1}$ = (s,b,a,t). The labeling results of the earliest arrival flow algorithm are

For δ = 0, no labeling
For δ = 1, no labeling

4.5 Dynamic Flow Algorithms

For $\delta = 2$, no labeling,

For $\delta = 3$, dispatch one unit along $P_{3,1}$ at time $\delta - 3 = 0$.
This creates the following labels (see Fig. 4.15):

0 on arc (s,a)
1 on arc (a,b)
2 on arc (b,t)

For $\delta = 4$, dispatch one unit along chain $P_{3,1}$ at time $\delta - 3 = 1$.
This creates the following labels:

1 on arc (s,a)
2 on arc (a,b)
3 on arc (b,t)

For $\delta = 5$, dispatch one unit along chain $P_{3,1}$ at time $\delta - 3 = 2$.
This creates the following labels:

2 on arc (s,a)
3 on arc (a,b)
4 on arc (b,t)

Dispatch one unit along chain $P_{5,1}$ at time $\delta - 5 = 0$. This creates the following labels:

0 on arc (s,b)
2 off arc (a,b)
2 on arc (a,t)

For $\delta = 6$, dispatch one unit along chain $P_{3,1}$ at time $\delta - 3 = 3$.
This creates the following labels:

3 on arc (s,a)
4 on arc (a,b)
5 on arc (b,t)

Dispatch one unit along chain $P_{5,1}$ at time $\delta - 5 = 1$. This creates the following labels:

1 on arc (s,b)
3 off arc (a,b)
3 on arc (a,t)

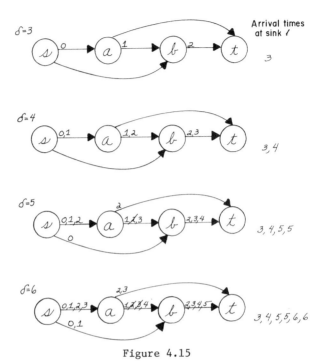

Figure 4.15

Example of a Earliest Arrival Flow Algorithm
[All arc capacities equal 1;
$a(s,a) = a(a,b) = a(b,t) = 1$,
$a(s,b) = a(a,t) = 3$.]

The resulting set of labels corresponds to an earliest arrival flow the 6 time periods.

Before presenting the proof of the earliest arrival flow algorithm, three lemmas are needed:

LEMMA 4.1. Consider the sequence of all flow augmenting chains generated by the minimum cost flow algorithm. Suppose that two chains P' and P''' both contain arc (x,y) as a forward arc. Then there exists a flow augmenting chain P" between P' and P''' in this sequence that contains arc (x,y) as a backward arc.

Proof: Recall that the capacity of arc (x,y) is one unit. Without loss of generality, assume that P' precedes P''' in the

4.5 Dynamic Flow Algorithms

sequence of flow augmenting chains generated by the minimum cost flow algorithm. Since arc (x,y) is a forward arc in P', arc (x,y) is saturated when chain P' augments flow. A similar situation occurs when P''' augments flow. Hence, between these two flow augmentations, a flow unit must be removed from arc (x,y). Hence, arc (x,y) must be a backward arc in some flow augmenting chain P''. Q.E.D.

LEMMA 4.2. Suppose a flow unit travels along flow augmenting chain $P_{i,j}$ from s to t spending a(x,y) time periods in each forward arc (x,y) and -a(x,y) time periods in each backward arc (x,y). Then, the total travel time from s to t is i time periods.

Proof: When flow augmenting chain $P_{i,j}$ is generated by the minimum cost flow algorithm during its i-th iteration p(s) = 0, p(t) = i and p(y) - p(x) = a(x,y) for all arcs (x,y) in flow augmenting chain $P_{i,j}$, and the lemma follows. Q.E.D.

LEMMA 4.3. Let $p_i(x)$ denote the value assigned to p(x) during the i-th iteration of the minimum cost flow algorithm. Then, for $i \leq j$,

$$p_i(x) \leq p_j \leq p_i(x) + j - i$$

Proof: At the beginning of each iteration of the minimum cost flow algorithm, each vertex number either increases by +1 or remains unchanged. Hence, $p_{i+1}(x) = p_i(x)$ or $p_{i+1}(x) = p_i(x) + 1$, and the lemma follows. Q.E.D.

Proof: To prove the algorithm, we must prove (a)

(a) That the labels generated by the algorithm correspond to a flow
(b) That this flow is an earliest arrival flow for p time periods
(c) That the algorithm terminates after a finite number of steps.

Proof of (a): To show that the labels produced by the algorithm correspond to a flow, it suffices to prove two claims:

(1) The algorithm never indicates the removal of a label that does not exist already
(2) The algorithm never duplicates a label on an arc.

To prove the first claim, suppose that arc (x,y) is a backward arc in chain $P_{m,n}$. Consequently, arc (x,y) must be a forward arc in some chain that precedes $P_{m,n}$. Let the last such augmenting chain preceding $P_{m,n}$ be denoted by $P_{i,j}$. It follows that $i \leq m$.

According to the instructions of the earliest arrival flow algorithm, chain $P_{i,j}$ labels arc (x,y) with $p_i(x)$, $p_i(x) + 1$, ..., $p_i(x) + m - i$, before chain $P_{m,n}$ removes any labels from arc (x,y). The chain $P_{m,n}$ removes the label $p_m(x)$ which is already present on arc (x,y) by Lemma 4.3. Moreover, before chain $P_{m,n}$ is required to remove another label from arc (x,y), chain $P_{i,j}$ has placed another label on arc (x,y). Since the additional labels are successive integers and since the labels to be removed are successive integers. it follows that any label to be removed from arc (x,y) has already been placed on arc (x,y). This proves the first claim.

To prove the second claim, suppose that arc (x,y) has received several identical labels. These labels must have been generated by distinct flow augmenting chains that contain arc (x,y) as a forward arc. Let $P_{i,j}$ and $P_{m,n}$ denote any two such flow augmenting chains. By Lemma 1, there exists a flow augmenting chain $P_{k,\ell}$ sequenced between $P_{i,j}$ and $P_{m,n}$ that contains arc (x,y) as a backward arc. Chain $P_{k,\ell}$ removes labels $p_k(x)$, $p_k(x),+ 1$, ..., $p_k(x) + m - k$ from arc (x,y) before chain $P_{m,n}$ adds any labels to arc (x,y). The first label that chain $P_{m,n}$ generates on arc (x,y) is $p_m(x)$, which by Lemma 3 has already been removed. Similarly during each successive iteration of the algorithm, chain $P_{k,\ell}$ removes a label before chain $P_{m,n}$ places the same label on arc (x,y). This completes the proof of the second claim.

Proof of (b): We shall now demonstrate that the algorithm does in fact produce an earliest arrival flow. This is accomplished by an inductive argument. Clearly, the algorithm constructs an earliest arrival flow for $p = 0$. Suppose that the algorithm constructs an earliest arrival flow for $p = p_0 - 1$. It remains to show that the algorithm constructs an earliest arrival flow for $p = p_0$.

4.5 Dynamic Flow Algorithms

The flow generated by the algorithm for p_0 time periods consists of the flow generated by the algorithm for $p_0 - 1$ time periods plus some additional flow units arriving at the sink at time period p_0. By Lemma 2, for each flow augmenting chain $P_{0,1}, \ldots, P_{0,r_0}, P_{1,1}, \ldots, P_{1,r_1}, \ldots, P_{p_0,1}, \ldots, P_{p_0,w_{p_0}}$ there is a flow unit arriving at the sink at time period p_0. This follows since each of these flow augmenting chains contributes one flow unit to flow F_{p_0}. Hence, the total number of flow augmenting chains in the above sequence equals V_{p_0}.

By the induction hypothesis, the flow units arriving at the sink by time period $p_0 - 1$ in the flow generated by the algorithm for $p = p_0$ constitute an earliest arrival flow for $p_0 - 1$ time periods and hence also a maximum dynamic flow for $p_0 - 1$ time periods. By equation (49), it follows that the flow generated by the algorithm is not only a maximum dynamic flow for p_0 time periods but also an earliest arrival flow for p_0 time periods since V_{p_0} flow units arrive at t at time period p_0.

Proof of (c): The algorithm terminates in a finite number of steps since labels have to be placed or removed along only a finite number of flow augmenting chains. (There are only a finite number of flow augmenting chains, otherwise the minimum cost flow algorithm would not terminate after a finite number of steps.) Q.E.D.

In some situations, it might be preferable to have flow units arrive at the sink as late as possible rather than as early as possible. For example, if each flow unit represents a check drawn on your account, you might prefer that they arrive at the payee (sink) as late as possible so that your checking balance would be as large as possible. Or the flow units might represent manufactured goods whose storage is very expensive in which case you would prefer not to have them in storage (i.e., at the sink) until as late as possible. For situations such as these, we would be interested in a maximum dynamic flow in which the flow units arrive at the sink as late as possible rather than as early as possible.

A *latest arrival flow for p time periods* is any maximum dynamic flow for p time periods in which as many flow units as possible arrive during the last time period, as many flow units as possible arrive during the last two time periods, etc.

Does there always exist a latest arrival flow for p time periods? Yes. The existence of a latest arrival flow for p time periods can be proved by construction in the same way that the existence of an earliest arrival flow for p time periods was proved by construction. The only difference in the proof is that now units are first sent into sink t_p, then into sink t_{p-1}, ...

Needless to say, the construction of the latest arrival flow for p time periods suggested above is hardly efficient when the graph has many arcs or vertices or when p is large. Even for the special case when all arc capacities are stationary, there does not appear to be an efficient algorithm for constructing a latest arrival flow, unlike the case for the earliest arrival flow which could efficiently be generated by the earliest arrival flow algorithm. The reason for this is that a latest arrival flow necessitates that flow units leave the source and linger or holdover in the network as much as possible before arriving at the sink.

Up to now, we have discussed only the pattern of arrivals of flow units at the sink and ignored the pattern of departures of flow units from the source. There are, of course, many situations in which the departure pattern at the source is important. For example, if you are shipping manufactured items out of your warehouse, you might prefer to dispatch them into the distribution network as soon as possible in order to minimize the storage costs at your warehouse. On the other hand, if the flow units represent school boys returning to school (sink) after a summer holiday at home (source), it might be preferable for the flow units to remain at the source as long as possible (if the boys' sentiments are taken into account).

In a similar way, we can define an *earliest departure flow for p time periods* and a *latest departure flow for p time periods*. The existence of each of these flows can be proved constructively in a

4.5 Dynamic Flow Algorithms

way similar to the proof for the earliest arrival flow by substituting sources for sinks in the proof.

As was the case for the earliest (and latest) arrival flow, the construction suggested by the existence proof is not efficient computationally but seems to be the best method available.

Fortunately, however, for the special case when arc capacities are stationary through time, there is an efficient way to construct a latest departure flow. This construction is shown below:

Let $G = (X, A)$ be any graph. Define the *inverse graph* G^{-1} of graph G as the graph with vertex set X and arc set

$$A^{-1} = \{(y,x): (x,y) \in A\}$$

Thus graph $G^{-1} = (X, A^{-1})$ is simply graph G with its arc directions reversed. Let the capacity and traverse time of each arc in Γ^{-1} equal the capacity and traverse time for the corresponding arc in Γ.

Any flow from s to t in graph G corresponds to a unique flow from t to s in graph G^{-1} and viceversa by simply reversing the route taken by each flow unit.

LEMMA 4.4. Suppose there exists a maximum dynamic flow F for p time periods in graph G in which x_i flow units depart from the source during time period i and y_i flow units arrive at the sink during time period i, for $i = 0, 1, \ldots, p$. Then there exists a maximum dynamic flow for p time periods in graph G^{-1} in which x_i units arrive at the sink during time period $p - i$ and y_i units depart from the source during time period $p - i$, for $i = 0, 1, \ldots, p$.

Proof: The flow in graph G^{-1} that corresponds to flow F in graph G has the desired properties. Note that the flow units in graph G^{-1} travel backwards in time, and hence their departure and arrival times must be counted downward from p rather than from time zero. Q.E.D.

It follows from Lemma 4.4 that an earliest arrival flow in G for p time periods corresponds to a latest departure flow in graph G^{-1} for p time periods. Moreover, since $(G^{-1})^{-1} = G$, an earliest arrival flow in graph G^{-1} for p time periods corresponds to a latest departure flow in graph $(G^{-1})^{-1} = G$ for p time periods.

A latest departure flow for p time periods for graph G can be constructed as follows:

(1) Construct graph G^{-1} by reversing all arcs in G
(2) Determine an earliest arrival flow for p time periods for graph G^{-1} using the earliest arrival flow algorithm
(3) Determine the corresponding flow in graph G. This flow is a latest departure flow for p time periods for graph G.

Hence, the earliest arrival flow algorithm can also be used to generate a latest departure flow.

Now suppose that an earliest arrival flow for graph G for p time periods has already been found by the earliest arrival flow algorithm. Is it necessary to perform the earliest arrival flow algorithm again (this second time on graph G^{-1}) to determine the latest departure flow for graph G for p time periods? The following theorem answers this question negatively:

THEOREM 4.2. Suppose that the earliest arrival flow for graph G for p time periods consists of y_i flow units arriving at the sink during time period i, for i = 0, 1, ..., p. Then, the latest departure flow for graph G for p time periods consists of y_i units departing from the source during time period p - i, for i = 0, 1, ..., p.

Proof: We shall prove this result by contradiction. Suppose that in any latest departure flow for graph G for p time periods that Z, $Z \neq \sum_{j=0}^{j=i} y_j$, flow units depart from the source during the last i time periods. (This contradicts the theorem.) Then, from Lemma 4.4, there exists a maximum dynamic flow in graph G^{-1} for p time periods in which Z flow units arrive during the first i time periods at the sink. However, the maximum number of flow units that can be sent from source to sink in the last i time periods in graph G is equal to the maximum number of flow units that can be sent from the source to the sink in the first i time periods in graph G^{-1}. Thus, $Z = \sum_{j=0}^{j=i} y_j$, which is a contradiction. Q.E.D.

4.5 Dynamic Flow Algorithm

Hence, we can conclude that the earliest arrival and latest departure schedules for p time periods are symmetric, i.e., the number of flow units arriving at the sink at time period i in an earliest arrival flow equals the number of flow units leaving the source at time period p - i in a latest departure flow.

Happily, the earliest arrival flow constructed by the earliest arrival flow algorithm possesses another important property:

THEOREM 4.3. The earliest arrival flow constructed by the earliest arrival flow algorithm is also a latest departure flow.

Proof: Let $P_{i,j}$ denote, as before, the j-th flow augmenting chain found during the i-th iteration of the minimum cost flow algorithm. The earliest arrival flow algorithm for p time periods will send one flow unit along chain $P_{i,j}$ from the source to the sink at times 0, i, ..., p - i by Lemma 2. These flow units arrive at the sink at times i, i + 1, ..., p by Lemma 2. Since this is true for every flow augmenting chain $P_{i,j}$, for each flow unit arriving at the sink at time p - k there is a flow unit departing from the source at time k. Hence, the earliest arrival flow constructed by the earliest arrival flow algorithm has a latest departure schedule. Q.E.D.

For example, consider the earliest arrival flow generated by the earliest arrival flow algorithm in Fig. 4.15. The departure schedule of this flow is

Time period	0	1	2	3	4	5	6
Number of units	2	2	1	1	0	0	0

The arrival schedule for this flow is

Time period	0	1	2	3	4	5	6
Number of units	0	0	0	1	1	2	2

Notice the reverse symmetry between the arrival schedule and the departure schedule.

Unfortunately, the same reverse symmetry does not hold between latest arrival flows and earliest departure flows. To demonstrate this, consider the graph in Fig. 4.16. For p = 5, a maximum of 4 flow units can be sent from the source to the sink. The earliest departure schedule is by inspection:

Time period	0	1	2	3	4	5
Number of units	1	1	1	1	0	0

The latest arrival schedule is by inspection:

Time period	0	1	2	3	4	5
Number of units	0	0	0	1	1	2

Figure 4.16

Counterexample

(All arc capacities equal 1. Arc transit times are indicated.)

Consider a set S = {a,b,c, ...}. We say that there is a *lexicographic preference* a, b, c, ... on set S if one unit of item a is preferred to any number of units of the remaining items b, c, ..., and one unit of item b is preferred to any number of units of the remaining items c, ..., etc. An earliest arrival flow is called a *lexicographic flow* since it places a lexicographic preference t_0, t_1, t_2, ..., t_p on the sinks to be used by arriving flow units. Similarly, the latest arrival flow, earliest departure flow and latest departure flows are also lexigographic flows since there is for each of these flows some lexicographic preference on which sources or sinks the flow units are to use.

We shall conclude this section with a result which shows that for maximum flows any departure schedule at the source is compatible with any arrival schedule at the sink.

4.5 Dynamic Flow Algorithm

THEOREM 4.4. Let F' and F'' be any two maximum flows on graph G. Then, there exists a maximum flow on graph G that simultaneously has the same departure schedule at the source as flow F' and the same arrival schedule at the sink as flow F''.

Proof: Select any minimum cut separating the source from the sink in graph G. Since both F' and F'' are maximum flows, both saturate this cut. Construct a hybrid maximum flow from F' and F'' that is identical to flow F' on the source side of this cut and is identical to flow F'' on the sink side of this cut. This hybrid flow has the departure schedule identical to flow F' and an arrival schedule identical to flow F''. Thus, the hybrid flow satisfies the theorem. Q.E.D.

LEMMA 4.5. The following dynamic flows exist:

1. Earliest departure and earliest arrival
2. Earliest departure and latest arrival
3. Latest departure and earliest arrival
4. Latest departure and latest arrival

Proof: The proof is achieved by applying Theorem 4.4 to graph G_p. Q.E.D.

As shown in Theorem 4.3, the earliest arrival flow algorithm constructs an earliest arrival and latest departure flow, i.e., item 3 in Lemma 4.5. The three remaining flows, items 1, 2, and 4, could be constructed by splicing together two flows as suggested by the proof of Theorem 4.4. Unfortunately, finding more efficient methods of constructing the other three flows seems to be an open question.

4.6 FLOWS WITH GAINS

Previously, whenever a flow unit entered an arc, the same flow unit exited from the arc unchanged. Flow units were neither created nor destroyed as they traveled through an arc. In this section, we shall no longer assume that flow units are unchanged as they traverse an arc. Instead, we will allow the number of flow units traveling through an arc to increase or decrease. Specifically, we shall assume that if f(x,y) flow units enter arc (x,y) at vertex x, then k(x,y) f(x,y) flow units exit arc (x,y) at vertex y, for all arcs (x,y). Each flow unit traversing arc (x,y) can be regarded as being multiplied by k(x,y). The quantity k(x,y) is called the *gain* or *gain factor* of arc (x,y).

EXAMPLE 1. A plant nursery must dispatch a shipment of plants to its distributors. On some of these routes, the plants experience a high fatality rate due to improper climatic conditions. On other routes, with more suitable climatic conditions, the plants generally experience a significant growth during shipment.

The routes with plant fatalities can be regarded as arcs with gain factors that are less than one; the routes with plant growth can be regarded as arcs with gain factors that are greater than one.

EXAMPLE 2. A corporate financial analyst must decide how to ration the corporation's investment funds between competing investments. How can he develop a network with gains to aid his investment decision problem?

The analyst can regard each investment possibility as an arc going from the vertex corresponding to its starting date to the vertex corresponding to its termination date. If, say, an investment pays 8%, then a gain of 1.08 can be attached to the arc representing this investment. The capacity of each arc should equal the maximum amount that can be placed into the corresponding investment. If each dollar to be invested represents a flow unit, then the analyst's investment decision problem can be rephrased as the problem of sending as many flow units to the sink given a limited supply of flow units at the source of the graph generated by the investment possibilities. (The source is the vertex representing the current time and the sink is the vertex representing the date on which all investments must expire.)

If $k(x,y) > 1$, then flow is *increased* along arc (x,y). If $k(x,y) = 1$, then flow remains *unchanged* along arc (x,y). If $0 < k(x,y) < 1$, then flow is *reduced* along arc (x,y). If $k(x,y) = 0$, then flow is

4.6 Flows with Gains

eradicated by arc (x,y) and arc (x,y) can be regarded as a sink. In the discussion to follow we shall always assume that all arcs (x,y) with $k(x,y) = 0$ have been replaced by sinks. If $k(x,y) < 0$, then for each flow unit entering arc (x,y) at vertex x, $-k(x,y)$ flow units must arrive at vertex y; thus, arc (x,y) can be regarded as *creating a demand* for flow units.

The central problem of this section is the minimum cost flow with gains problem. As you might expect, this problem is the problem of finding a minimum cost way of dispatching a given number V of flow units from the source to the sink in a network in which the arcs have gain factors associated to them. Of course, if V flow units are discharged into the network at the source, then the number of flow units arriving at the sink need not equal V because of arc gains. (Note that in all previous flow problems, the number of units discharged into the network at the source always equaled the number that arrived at the sink.)

As before, arc (x,y) has associated with it an upper bound $c(x,y)$ and a lower bound $l(x,y)$ on the number of flow units that may enter it and a cost $a(x,y)$ for each unit that enters the arc. Again, let $f(x,y)$ denote the number of flow units that enter arc (x,y).

The minimum cost flow with gains problem can be written as follows:

Minimize

$$\sum_{(x,y)} a(x,y)\, f(x,y) \qquad (50)$$

such that

$$\sum_y f(x,y) - \sum_y k(y,x)\, f(y,x) = \begin{cases} V & \text{If } x = s \\ 0 & \text{If } x \neq s,\ x \neq t. \end{cases} \qquad (51)$$

$$l(x,y) \leq f(x,y) \leq c(x,y) \quad [\text{for all } (x,y)] \qquad (52)$$

Expression (50) represents the total cost of the flow. Equation (51) states that the net flow out of vertex s must equal V and the net flow out of every other vertex, except t, must equal zero. Relation (52) states that each arc flow must lie within the upper and lower bounds on the arc's capacity.

The minimum cost flow with gains problem, specified by relations (50) - (52) is a linear programming problem. Let $p(x)$ denote the dual variable corresponding to equation (51) for vertex x. Let $\gamma_1(x,y)$ denote the dual variable corresponding to the upper capacity constraint in relation (52) for arc (x,y). Let $\gamma_2(x,y)$ denote the dual variable corresponding to the lower capacity constraint in relation (52) for arc (x,y). With these definitions for the dual variables, the dual linear programming problem for the linear programming problem of relations (50) - (52) is

Maximize

$$Vp(s) - \sum_{(x,y)} c(x,y) \gamma_1(x,y) + \sum_{(x,y)} \ell(x,y) \gamma_2(x,y) \tag{53}$$

such that

$$p(x) - k(x,y) p(y) - \gamma_1(x,y) + \gamma_2(x,y) \leq a(x,y) \tag{54}$$

[for all arcs (x,y)]

$$\gamma_1(x,y) \geq 0 \quad \text{[for all arcs } (x,y)\text{]} \tag{55}$$

$$\gamma_2(x,y) \geq 0 \quad \text{[for all arcs } (x,y)\text{]} \tag{56}$$

$p(x)$ unrestricted (for all x) \hfill (57)

If the values for the dual variables $p(x)$, $x \in X$, are already given, then the remaining dual variables $\gamma_1(x,y)$ and $\gamma_2(x,y)$ must satisfy

$$-\gamma_1(x,y) + \gamma_2(x,y) \leq a(x,y) - p(x) + k(x,y) p(y) \triangleq \zeta(x,y)$$

[for all arcs (x,y)] \hfill (58)

For convenience, the right side of relation (58) is denoted by $\zeta(x,y)$.

The dual objective function (53) is maximized if $\gamma_1(x,y)$ and $\gamma_2(x,y)$ are chosen as follows:

If

$$\zeta(x,y) \geq 0,$$

set

$$\gamma_1(x,y) = 0 \quad \gamma_2(x,y) = \zeta(x,y)$$

4.6 Flows with Gains

If
$$\zeta(x,y) < 0$$
set
$$\gamma_1(x,y) = -\zeta(x,y) \quad \gamma_2(x,y) = 0 \tag{59}$$

This follows since $\gamma_1(x,y)$ and $\gamma_2(x,y)$ appear in only one dual constraint (54).

Thus, the values of $\gamma_1(x,y)$ and $\gamma_2(x,y)$ for all arcs (x,y) are imputed by the values of the dual variables $p(x)$. Thus, we need only seek an optimal set of values for the vertex dual variables $p(x)$.

The complementary slackness conditions generated by this primal-dual pair of linear programming problems are

$$\gamma_1(x,y) > 0 \Rightarrow f(x,y) = c(x,y) \tag{60}$$
$$\gamma_2(x,y) > 0 \Rightarrow f(x,y) = 1(x,y) \tag{61}$$

Since the values of $\gamma_1(x,y)$ and $\gamma_2(x,y)$ are determined by the value of $\zeta(x,y)$, the complementary slackness conditions, (60) and (61), can be restated in terms of $\zeta(x,y)$ as follows:

$$\zeta(x,y) < 0 \Rightarrow f(x,y) = c(x,y) \tag{62}$$
$$\zeta(x,y) > 0 \Rightarrow f(x,y) = 1(x,y) \tag{63}$$

Hence, to solve the minimum cost flow with gains problem, we need only find a set of feasible flow values $f(x,y)$ and a set of dual vertex variable values $p(x)$ such that the complementary slackness conditions, (62) and (63), are satisfied for all arcs (x,y).

Consider the cycle in Fig. 4.17 in isolation from the remainder of its graph. If one additional flow unit arrives at vertex x and enters arc (x,y), then during its passage across arc (x,y), it increases to two flow units at vertex y, and becomes 2/3 of a flow unit at vertex z, and finally becomes 2 flow units upon its return to vertex x. Thus, each additional flow unit that travels clockwise around this cycle becomes two flow units. This cycle is called a *clockwise generating cycle*. In general, a cycle C is called a clockwise generating cycle if

$$k_C \triangleq \frac{\prod_{\substack{(x,y) \\ \text{forward} \\ \text{arc}}} k(x,y)}{\prod_{\substack{(x,y) \\ \text{backward} \\ \text{arc}}} k(x,y)} > 1 \qquad (64)$$

when cycle C is traversed in the clockwise direction. Similarly, we can define a *counterclockwise generating cycle* as any cycle for which inequality (64) is satisfied when the cycle is traversed in the counterclockwise direction. The generating cycles (i.e., clockwise generating cycles and counterclockwise generating cycles) are important because they have the ability to create additional flow units.

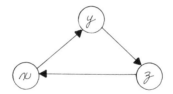

Figure 4.17
Clockwise Generating Cycle
[$k(x,y) = 2$, $k(y,z) = 1/3$, $k(z,x) = 3$.]

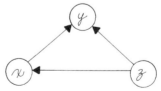

Figure 4.18
Clockwise Absorbing Cycle
[$k(x,y) = 1$, $k(z,y) = -1/2$, $k(z,x) = 1/4$.]

Now, consider the cycle in Fig. 4.18. If an additional flow unit arrives at vertex x and enters arc (x,y) and traverses this cycle in the clockwise direction, then an additional 1/2 flow unit

4.6 Flows with Gains

will be required at vertex x. This follows from the fact that the unit traversing arc (x,y) creates an additional unit at vertex y. This additional flow unit arriving at vertex y requires the flow in arc (z,y) to increase by 2 units. Hence, because of the increase of 2 units in arc (z,y), an additional 1/2 unit is required at vertex x. Thus, if 1 1/2 flow unit arrive at vertex x, one flow unit can be sent clockwise around this cycle and the remaining 1/2 flow unit can be used to supply the additional need at vertex x. Thus, this cycle absorbs flow units and is called a *clockwise absorbing cycle*. In general a cycle C is called a clockwise absorbing cycle if

$$k_C \triangleq \frac{\prod_{\substack{(x,y) \\ \text{forward} \\ \text{arc}}} k(x,y)}{\prod_{\substack{(x,y) \\ \text{backward} \\ \text{arc}}} k(x,y)} < 1 \tag{65}$$

when cycle is traversed in the clockwise direction. Similarly, we can define a *counterclockwise absorbing cycle* as any cycle for which inequality (65) is satisfied when the cycle is traversed in the counterclockwise direction.

Because a graph might contain generating and/or absorbing cycles, we cannot be certain that every flow unit leaving the source ultimately arrives at the sink (it could be absorbed by an absorbing cycle) or that every flow unit arriving at the sink initially came from the source (it could have been generated by a generating cycle).

Note that if k_C is calculated in the clockwise direction, then $1/k_C$ equals the value of k_C for the counterclockwise direction. This follows since changing the direction of travel around the cycle changes backward arcs to forward arcs and forward arcs to backward arcs. Hence, the numerator and denominator in k_C are switched.

If a graph contains no absorbing or generating cycles, that is, $k_C = 1$ for all cycles C for both clockwise and counterclockwise directions, then the minimum cost flow with gains problem can be converted into a pure (i.e., without gains) minimum cost flow problem which can be solved by the Out-of-Kilter Algorithm. We shall now show how to convert the minimum cost flow with gains problem into a minimum cost flow problem.

Consider the conservation of flow requirements (51) and the flow capacity constraints (52) for the minimum cost flow with gains problem. Let E denote the coefficient matrix of the left side of these constraints. If a row of E is multiplied by a nonzero constant, the set of feasible solutions to this set of simultaneous linear inequalities remains unchanged.

Suppose that the column of E corresponding to arc (x,y) is multiplied by a nonzero constant c. Then, in the original set of inequalities cf(x,y) appears wherever f(x,y) formerly appeared. If l(x,y) and c(x,y) are also multiplied by c, then cf(x,y) can be replaced by a new variable f'(x,y) and the constraints remain those of the original problem.

Can we find row and column multipliers such that after the redefinition of variables as defined above, the new flow problem will be a pure network flow problem? The following theorems due to Truemper (1976) provide an affirmative answer to this question:

THEOREM 4.5. A flow with gains problem can be converted into a flow without gains problem if, and only if, there exist a vertex number m(x) for each vertex x such that

$$\frac{m(x) \ k(x,y)}{m(y)} = 1 \qquad (66)$$

for all arcs (x,y) in the graph.

Proof: Suppose there exist vertex numbers m(x) such that equation (66) is satisfied for all arcs (x,y) in the graph. Multiply the conservation of flow equation (51) for vertex x by 1/m(x) for each vertex x. Let f'(x,y) = f(x,y)/m(x). for all arcs (x,y). Rewrite equations (51) and (52) in terms of the f' variables. After doing this, the resulting equations in the f' variables are in the form of relations (24) - (26), which is a minimum cost flow (without gains) problem.

4.6 Flows with Gains

To prove the converse, suppose that the flow with gains problem can be converted into a flow without gains problem. Then, there must exist a multiplier for each row in E that converts the corresponding equation into the constraint for a flow without gains problem. Let $1/m(x)$ denote the multiplier for constraint (51) for vertex x.

In a flow without gains problem, the coefficient of each variable $f'(x,y)$ in the conservation of flow constraint (25) for vertex x is +1; consequently, $f'(x,y) = f(x,y)/m(x)$. In a flow without gains problem, the coefficient of each variable $f'(y,x)$ in the conservation of flow constraint (25) for vertex x is -1: Consequently, $-f'(y,x) = -k(y,x) f(y,x)/m(x) = -k(y,x) m(y) f'(y,x)/m(x)$. Hence, $1 = k(y,x) m(y)/m(x)$ for all arcs (y,x) which is condition (66). Q.E.D.

Thus, a flow with gains problem can be converted into a pure flow problem if there exist vertex numbers satisfying the conditions prescribed in (66). When do such vertex numbers exist? How can they be calculated? Read on.

THEOREM 4.6. Vertex numbers satisfying Theorem 4.5 exist, if, and only if, the graph contains no absorbing and no generating cycles, that is, $k_C = 1$ for all cycles C.

Proof: Suppose that a set of vertex numbers $m(x)$ exist such that $m(x,y) = k(x,y)m(x)/m(y) = 1$ for all arcs (x,y). Select any cycle C. It follows that

$$k_C = \frac{\prod\limits_{\substack{(x,y) \\ \text{forward}}} k(x,y)}{\prod\limits_{\substack{(x,y) \\ \text{backward}}} k(x,y)} = \frac{\prod\limits_{\substack{(x,y) \\ \text{forward}}} \frac{m(x)k(x,y)}{m(y)}}{\prod\limits_{\substack{(x,y) \\ \text{backward}}} \frac{m(x)k(x,y)}{m(y)}}$$

$$= \frac{\prod\limits_{\substack{(x,y) \\ \text{forward}}} m(x,y)}{\prod\limits_{\substack{(x,y) \\ \text{backward}}} m(x,y)} = 1 \qquad (70)$$

Hence, if all m(x,y) equal 1, no absorbing or generating cycle can exist in the graph.

Conversely, suppose that no generating or absorbing cycle exists in the graph. Select any spanning tree T of the graph. Let m(s) = 1, and let m(x,y) = 1 for all arcs (x,y) in T. In the obvious way, fan out along the arcs of the tree calculating a value for m(x) for each vertex x. These vertex numbers are uniquely determined.

Select any arc (x,y) not in T. This arc forms a unique cycle C with the arcs in T. By (70), it follows that m(x,y) = 1 since m(i,j) = 1 for all arcs (i,j) in T. Thus, m(x,y) = 1 for all arcs (x,y) not in T. Q.E.D.

Thus, if a network possesses no absorbing or generating cycles, then the minimum cost flow with gains problem can be transformed into a minimum cost flow without gains problem, which can be solved optimally by the out-of-kilter algorithm or in the case of all lower capacities equalling zero by the minimum cost flow algorithm. This transformation is effected by multiplying the conservation of flow constraint (51) for vertex x by 1/m(x) where m(x) is determined as in the proof of Theorem 4.6.

If the network possesses absorbing or generating cycles, the flow problem becomes more complicated. In this section, we shall present an algorithm, Jewell (1962), for solving the minimum cost flow with gains problem. This algorithm is called the *flow with gains algorithm*. It consists of three basic steps:

Step 1 (Initialization Step): This step finds flow values f(x,y) and dual variable vertex number values p(x) such that the complementary slackness conditions (62) - (63), are satisfied and that the flow satisfies all feasibility requirements, (51)-(52), except that possibly less than V flow units are dispatched from the source. Hence, Step 1 finds a solution that satisfies all primal, dual and complementary slackness requirements, except the required output at the source. (As will be seen later, Step 1 will be expedited by constructing a graph that is a slight enlargement of the original graph.)

4.6 Flows with Gains

Step 2 (Flow Increment Step): In this step, the flow out of the source is increased while retaining all the primal, dual and complementary slackness requirements described in Step 1.

Step 3 (Dual Variable Change Step): This step describes how to change the values of the dual variables p(x) so that even more flow units can be discharged out of the source into the network.

The algorithm consists of performing Step 1 to determine an initial flow. Step 2 is performed next to send more flow units out of the source, while maintaining all complementary slackness conditions. Once Step 2 cannot discharge any more flow units from the source into the network while maintaining all complementary slackness conditions, Step 3 is performed to change the dual variable values so that even more flow units can be discharged into the network. The algorithm returns to Step 2. Once Step 2 ceases to discharge any more flow units into the network, Step 3 is repeated, etc., until all V flow units have been discharged into the network from the source. If no feasible flow that discharges V units into the network exists, the algorithm will discover this during an iteration of Step 3 and terminate.

With all this as motivation, we are now prepared to state formally the *flow with gains algorithm*.

Flow With Gains Algorithm

Step 1 (Initialization): This step shows how to find a set of flow values $f(x,y)$ and dual variable values $p(x)$ for a network equivalent to our original network such that all feasibility conditions (51) - (52) and complementary slackness conditions (62) - (63) are satisfied except that the net flow out of source s is less than or equal to V, the required flow out of source s.

Arbitrarily select any values for the dual variables $p(x)$, $x \in X$. Consult the complementary slackness conditions

$$\zeta(x,y) = a(x,y) - p(x) + k(x,y)p(y) < 0 \Rightarrow f(x,y) = c(x,y)$$

(62)

$$\zeta(x,y) = a(x,y) - p(x) + k(x,y)p(y) > 0 \Rightarrow f(x,y) = 1(x,y)$$
(63)

to determine which values for f(x,y) are compatible with the complementary slackness conditions for each arc (x,y). Select any compatible value for f(x,y) for each arc (x,y).

Next, determine the net surplus flow V(x) out of each vertex x, where

$$V(s) = \sum_y f(x,y) - \sum_y k(y,x)f(y,x) - V$$
$$V(x) = \sum_y f(x,y) - \sum_y k(y,x)f(y,x)$$
(71)

If V(x) = 0 for all x, then the current solution is an optimal solution since it is a feasible flow that satisfies all complementary slackness conditions for the chosen values of p(x). Of course, we usually are not so fortunate.

Create a new network from the original network by adding a vertex S and an arc (S,x) from S to each vertex x with V(x) ≠ 0. Let c(S,x) = V(x) for each arc (S,x).

If
    V(x) < 0
let
    k(S,x) = +1
If
    V(x) > 0
let
    k(S,x) = -1

Lastly, let a(S,x) = 0 for all arcs (S,x).

Let the flow in each newly created arc (S,x) equal zero and let the flow in all other arcs remain as before. Clearly, flow is not conserved at all intermediate vertices of the new network since at least one intermediate vertex x has V(x) ≠ 0. However, if enough flow units can be sent out of vertex S and absorbed into the new network so that each arc (S,x) is saturated, then this new (non-feasible) flow in the new network will correspond to a feasible flow

4.6 Flows with Gains

in the original network. This follows since arc (S,x) will supply vertex x with exactly enough flow units to counteract the net surplus out of vertex x.

Moreover, a minimum cost flow in the new network that saturates all arcs (S,x) corresponds to a minimum cost flow in the original network since all a(S,x) = 0. Hence, we can find an optimal solution for the original network by searching for an optimal solution for the new network.

The linear programming formulation of this minimum cost flow with gains problem on the new network is like the linear programming formulation of the minimum cost flow with gains problem on the original network, relations (50)-(52), except that now vertex S is the source vertex and the right side of the conservation of flow equation (51) for vertex x should now be V(x) instead of zero.

The complementary slackness conditions for the new network are the same as for the original network.

If we retain the same values as above for p(x) for each vertex x in the original network and let p(S) be any large negative number, the complementary slackness conditions will be satisfied for the new network.

Hence, the original selection of flow values and dual variable values generates a set of values for the flow variables and dual variables for the new network that satisfy feasibility and complementary slackness conditions. All is done, except more units must be sent into the network from vertex S. Proceed to Step 2.

Step 2 (Flow Augmentation): This step shows how to increase the flow out of the source S as much as possible without losing any complementary slackness or changing any dual variable value.

For each arc (x,y), determine if the complementary slackness conditions (62)-(63) will allow f(x,y) to be increased or decreased. Let I denote the set of increasable arcs, and let R denote the set of reducible arcs. Only the arcs in I and R will be considered in this step.

To determine if additional flow units can be dispatched from S using only arcs in I and R, we shall successively grow a tree of arcs rooted at the source, just as in the maximum flow algorithm. These arcs will be called *colored arcs*. If a vertex is reached by this tree, then additional flow units can be dispatched from the source to this vertex via the arcs of the tree. Whenever an arc incident to a vertex x is added to the tree (colored), vertex x will receive a label f(x) denoting the number of flow units that would arrive at vertex x for each flow unit sent to it from the source.

If the sink is labeled, then a flow augmenting chain from the source to the sink has been discovered. As much flow as possible is dispatched from the source along this flow augmenting chain to the sink.

If an arc that forms a cycle with the colored arcs can also be colored, then we must check if this cycle can absorb flow units sent out from the source. If so, we dispatch as much flow as possible from the source to be absorbed by this cycle. If this cycle cannot absorb flow units dispatched from the source, then some of the colored arcs in this cycle are uncolored, and the coloring process continues.

Now for the details of the coloring and labeling procedure: Initially all arcs are uncolored and all vertices are unlabeled. Label the source S with the label f(S) = +1. Perform the following coloring and labeling operations:

Uncolored arc (x,y) can be colored if one of the following conditions is met:

1. x is labeled, $f(x) > 0$, and $(x,y) \in I$
2. x is labeled, $f(x) < 0$, and $(x,y) \in R$
3. y is labeled, $f(y) > 0$, and either $(x,y) \in R$, $k(x,y) > 0$ or $(x,y) \in I$, $k(x,y) < 0$
4. y is labeled, $f(y) < 0$, and either $(x,y) \in I$, $k(x,y) > 0$ or $(x,y) \in R$, $k(x,y) < 0$.

If arc (x,y) is colored because of items 1 or 2 then we say that arc (x,y) was colored from vertex x. In this case, label y with $f(y) - f(x)k(x,y)$.

4.6 Flows with Gains

If arc (x,y) is colored because of item 3 or 4, then we say that arc (x,y) was colored from vertex y. In this case, label x with $f(x) = f(y)/k(x,y)$.

Note that a vertex label $f(x)$ denotes the number of flow units that arrive at vertex x for each unit discharged at S along the chain of colored arcs that generated the label $f(x)$.

Continue this coloring and labeling procedure until

1. The sink is labeled
2. Some vertex receives two distinct labels
3. No more labeling or coloring is possible.

If item 3 occurs, go to Step 3.

If item 1 occurs, then a unique chain of arcs from the source to the sink has been colored. This is a flow augmenting chain from the source to the sink. Send as many units as possible along this chain, and return to the beginning of Step 2.

If item 2 occurs, then some vertex x has received two distinct labels, $f_1(x)$ and $f_2(x)$. Without loss of generality, suppose that $f_1(x) < f_2(x)$. Each label corresponds to a distinct (possibly partially overlapping) flow augmenting chain from S to x. If the labels $f_1(x)$ and $f_2(x)$ are of opposite signs, then a flow unit sent from S along the flow augmenting chain corresponding to the label $f_1(x)$ results in a demand of $f_1(x)$ flow units at vertex x. A flow unit sent from S along the flow augmenting chain corresponding to the label $f_2(x)$ results in $f_2(x)$ flow units arriving at x. Consequently, these two flow augmenting chains contain an absorbing cycle. Send as many flow units as possible from S along these two chains so that the flow units required at x by the first chain are supplied by the flow units arriving at x along the second chain. Return to the beginning of Step 2.

If $f_1(x)$ and $f_2(x)$ have the same sign, then we cannot be certain that we have discovered an absorbing cycle that can accept flow units discharged at S. Without loss of generality, suppose that $|f_1(x)| < |f_2(x)|$. If the label $f_2(x)$ is descendent from the label $f_1(x)$

[i.e., involves $f_1(x)$ in its computation), then a cycle C has been labeled from vertex x back to itself. Since, $|f_1(x)| < |f_2(x)|$, it follows that $k_C = f_1(x)/f_2(x) < 1$, and hence, cycle C is an absorbing cycle. Send as many flow units as possible from the source to be absorbed by cycle C, and return to the beginning of Step 2.

If label $f_2(x)$ is not descendent from label $f_1(x)$, then we have discovered two alternate flow augmenting chains from S to x but have not discovered any absorbing cycle. In this case, erase label $f_2(x)$ and all coloring and labels descendent from label $f_2(x)$. Continue the coloring and labeling procedure.

Step 3 (Dual Variable Change): This step shows how to define new values $p'(x)$ for the dual variables $p(x)$ that will satisfy the complementary slackness conditions. After the values of the dual variables are redefined, return to Step 2 to try to discharge more flow units from the source.

The details of redefining the dual variables are as follows: Let T denote the tree of arcs colored during the last iteration of Step 2. Let L denote the set of vertices labeled during the last iteration of Step 2. Set L consists of the endpoints of all arcs in T.

For each vertex x, define a variable $q(x)$ as follows:

$$q(x) = \begin{cases} \dfrac{1}{f(x)} & \text{If } x \in L \\ 0 & \text{If } x \notin L \end{cases} \qquad (72)$$

Define

$$\delta = \min\left\{\frac{\zeta(x,y)}{[q(x) - k(x,y)q(y)]}\right\}$$

where the minimization is taken over all arcs (x,y) for which the numerator and denominator have the same sign. If the denominator is zero for all arcs, then stop the algorithm since no feasible solution exists for the minimum cost flow with gains problem. Otherwise, define the new values $p'(x)$ for the dual variables as follows:

$$p'(x) = p(x) + \delta q(x) \quad \text{(for all x)} \qquad (73)$$

Return to Step 2.

4.6 Flows with Gains

Proof of the Flow with Gains Algorithm: We must show that the flow with gains algorithm terminates with an optimal solution or that no solution exists.

Step 1 transforms the original minimum cost flow with gains problem to a new network in which each source arc must be saturated. A minimum cost solution in the new network in which every source arc is saturated is equivalent to an optimal solution to the minimum cost flow with gains problem in the original network. Hence, we need only show that (a) the algorithm finds an optimal solution to the minimum cost flow with gains solution in the new network that saturates all source arcs, or (b) no feasible solution exists.

(a) To show that the algorithm terminates with an optimal solution in the new network, we need only show that complementary slackness is maintained at all times since the algorithm can only terminate with a flow that saturates all source arcs or else the algorithm claims no feasible solution exists.

The algorithm is initialized in Step 1 with a solution in which all complementary slackness conditions are satisfied. Moreover, all complementary slackness conditions are maintained throughout all flow changes in Step 2 since the flow changed in an arc are never permitted to violate any complementary slackness conditions.

Are all complementary slackness conditions maintained by Step 3? Consider any arc $(x,y) \in A$. After the vertex number change of Step 3, the new value $\zeta'(x,y)$ of $\zeta(x,y)$ becomes

$$a(x,y) - p'(x) + k(x,y)p'(y)$$
$$= a(x,y) - p(x) - \delta q(x) + k(x,y)[p(y) + \delta q(y)] \quad (74)$$
$$= \zeta(x,y) + \delta[-q(x) + k(x,y)q(y)]$$

Case 1. If arc $(x,y) \in T$, then $q(x) = k(x,y)q(y)$ and $\zeta'(x,y) = \zeta(x,y) + \delta[-q(x) + q(x)] = \zeta(x,y)$. Since the dual variable change of Step 3 causes no change in $\zeta(x,y)$, it follows that the complementary slackness conditions remain satisfied for arc (x,y).

Case 2. If arc $(x,y) \notin T$, $x \notin L$, $y \notin L$, then $q(x) = q(y) = 0$ and $\zeta'(x,y) = \zeta(x,y) + 0\delta = \zeta(x,y)$. Since the dual variable change of Step 3 causes no change in $\zeta(x,y)$, it follows that the complementary slackness conditions remain satisfied for arc (x,y).

Three more cases remain:

$(x,y) \notin T$, $x \in L$, $y \notin L$
$(x,y) \notin T$, $x \notin L$, $y \in L$
$(x,y) \notin T$, $x \in L$, $y \in L$.

It is left to the reader to verify that for each of these cases that the dual variable change maintains all complementary slackness conditions for arc (x,y). These verifications follow the same lines as above.

(b) Suppose that the algorithm terminates in Step 3 with δ undefined. We shall now show that no feasible flow exists.

Additional flow from the source can reach only the vertices in set L without violating any complementary slackness conditions imposed by the current choice of the dual variables $p(x)$. Can the dual variable values be changed so that set L can be increased by at least one member, thereby bringing us closer to labeling the sink or coloring an absorbing cycle? Disregarding the complementary slackness conditions, there are five ways an arc (x,y) can be colored:

1. $x \in L$, $y \notin L$, $f(x) > 0$, $f(x,y) < c(x,y)$
2. $x \in L$, $y \notin L$, $f(x) < 0$, $f(x,y) > l(x,y)$
3. $x \notin L$, $y \in L$, $f(y) > 0$, $f(x,y) > l(x,y)$
4. $x \notin L$, $y \in L$, $f(y) < 0$, $f(x,y) < c(x,y)$
5. $x \in L$, $y \in L$, $(x,y) \notin T$, and (x,y) forms an absorbing cycle with the colored arcs.

Careful examination of each of these cases shows that for each case $\zeta(x,y)$ and $q(x) - k(x,y)q(y)$ must have the same sign. If δ is undefined, then it follows that there is no arc (x,y) that is uncolored and for which $\zeta(x,y)$ and $q(x) - k(x,y)q(y)$ have the same sign. Hence, no arc is a candidate for coloring and flow units cannot reach any further out of the source. Hence, no feasible flow exists. Q.E.D.

4.6 Flows with Gains

The flow with gains algorithm and its proof as stated here avoided any mention of the number of steps required before termination. In fact, the algorithm as stated need not terminate in a finite number of steps. We shall now describe how to modify the flow with gains algorithm to insure that it will terminate finitely. First some definitions are needed.

A flow with gains is called *canonical* if no connected component of the set of intermediate arcs (arcs with flow strictly between the arc's lower and upper capacity) contains any of the following configurations:

1. A cycle C with $k_C = 1$
2. Two distinct but possibly overlapping cycles
3. The source and a cycle
4. The sink and a cycle
5. The source and the sink

Notice that if a connected component of intermediate arcs contains any of the above configurations, then the flow in the arcs in the component can be altered so that

(a) One or more arcs in the component becomes nonintermediate
(b) The net flow out of each vertex (except possibly the source and the sink) remains unchanged.

For example, if the intermediate arcs contain two cycles that overlap or are connected to each other via intermediate arcs (configuration 2), then flow could be increased around one of the cycles, and the surplus (or deficiency) created by the flow change around this cycle could be absorbed into the second cycle.

It is left to the reader to verify for himself that this flow change can always be made so that the net flow out of the source remains unchanged or is increased.

Hence, if a flow with gains is not canonical, it contains one of the above configurations and a flow change can be made to decrease the number of intermediate arcs without decreasing the net flow out of the source. If the new flow resulting from this flow change is also not canoncial, then this process can be repeated. After successive repetitions of this process a canonical flow will be generated without decreasing the net flow out of the source. Hence, any noncanonical flow can be converted into a canonical flow without decreasing the net flow out of the source.

THEOREM 4.7. There exist only a finite number of distinct canonical flows in graph G.

Proof: The number of distinct possibilities for the set M of intermediate arcs is finite since M is a subset of the finite arc set of graph G. The flow in each arc $(x,y) \notin M$ is either $l(x,y)$ or $c(x,y)$. Hence, there are only a finite number of possible sets of values for the flows of the arcs not in M.

For a given set M and a given set of flow values for the arcs not in M, we shall show that if an infinite number of distinct flows exist, then none of them are canonical. This will prove the theorem.

Some of the flow values of the arcs in M are uniquely determined by the conservation of flow requirements. Let M' denote the subset of M whose flow values are not uniquely determined by the conservation of flow requirements. Set M', if it is not empty, must contain a cycle or a chain from s to t. (Otherwise, the flow values of all members of M' could be successively determined starting at some vertex incident to only one member of M'.)

If set M' contains a chain from s to t, then none of the possible flows is canonical. If set M' contains a cycle C with $k_C = 1$ then none of the possible flows is canonical. Hence, if set M' is not empty, it must contain a cycle C with $k_C \neq 1$.

If an infinite number of different flows are possible, then an infinite number of different flows are possible around cycle C. Since $k_C \neq 1$, it follows that the surplus or deficiency caused by each of these flows in cycle C must be compensated for either by flow from the source or flow into the sink or by flow in some other cycle. Hence, the set I' must contain configuration 2, 3, or 4. Hence, none of the possible flows is canonical. Q.E.D.

4.6 Flows with Gains

With the above results in mind, we are now able to state formally the *finite termination modification* for the flow with gains algorithm.

Finite Termination Modification

Convert the initial flow generated by Step 1 of the flow with gains algorithm to a canonical flow without decreasing the net flow out of the source.

In Step 2 of the flow with gains algorithm, never color a nonintermediate arc if you can instead color an intermediate arc.

Proof: First we shall show that if Step 2 is started with a canonical flow, then the flow resulting from a flow change in Step 2 is also a canonical flow. Suppose that the new flow is not canonical. Then, the intermediate arcs of the new flow would contain at least one of the five configurations. Denote the (intermediate) arcs in this configuration by J. Let J' denote the arcs in J that were intermediate in the original flow. Let J" denote the arcs in J that were not intermediate in the original flow. When the original flow was changed to the new flow, Step 2 must have colored the arcs in J". This implies that, if the finite termination modification were used, that no arcs in J' were available for coloring each time an arc in J" was colored. However, careful examination of each of the five possible configurations will show that this nonavailability of the arcs in J' for coloring is impossible. This contradicts the instructions of the finite termination modification.

Hence, Step 2 of the algorithm will produce only a canonical flow from a canonical flow. Since each successive flow produced by Step 2 increases the net flow out of the source, and since there are by Theorem 3 only a finite number of distinct canonical flows, Step 2 will never generate the same canonical flow twice. Hence, Step 2 cannot be performed an infinite number of times, and the only possible way for the flow with gains algorithm not to terminate finitely is for Step 3 to be performed an infinite number of times.

If Step 3 were performed an infinite number of times, then there must be an infinite number of successive iterations of Step 3 between two flow changes since there are only a finite number of flow changes. However, after each iteration of Step 3 at least one more arc is colored, or δ is undefined and the algorithm terminates with the discovery that no feasible flow exists. Since there are only a finite number of arcs in the graph, Step 3 can be performed only a finite number of successive times on the same flow. Hence, the algorithm must terminate after a finite number of steps. Q.E.D.

Other Flows With Gains

So far, we have considered only the problem of finding a minimum cost flow with gains that discharges a given number of flow units from the source. Suppose that instead we were interested in finding a minimum cost flow that conveyed a given number of flow units to the sink. Could the flow with gains algorithm be used to solve this problem? The answer is yes. This is accomplished by using the inverse graph.

Recall that the *inverse graph* G^{-1} of graph $G = (x, A)$ was defined as the graph with vertex set X and arc set

$$A^{-1} = \{(y,x): (x,y) \in A\}$$

Let the cost of arc $(y,x) \in A^{-1}$ equal $a(x,y)/k(x,y)$. Let the gain of arc $(y,x) \in A^{-1}$ equal $1/k(x,y)$. Let $1(y,x) = 1(x,y)/|k(x,y)|$ and $c(y,x) = c(x,y)/|k(x,y)|$. Note that the inverse of G^{-1} is G.

Given any flow $f(x,y)$ for graph G, we can define the inverse flow $f^{-1}(y,x)$, $(y,x) \in A^{-1}$, as follows:

If $k(y,x) > 0$, then $f^{-1}(y,x) = f(x,y)k(x,y)$.

If $k(y,x) < 0$, let $c(x,y)$ units be supplied at vertex x,
let $k(x,y)c(x,y)$ units be supplied at vertex y,
and let $f^{-1}(y,x) = [c(x,y) - f(x,y)](-k(x,y))$.

Let $b(x)$ denote the total number of flow units supplied at vertex x.

Why this particular definition of the inverse flow? If in graph G, $f(x,y)$ flow units are removed from vertex x by arc (x,y) and $k(x,y)f(x,y)$ flow units are delivered to vertex y by arc (x,y), then

4.6 Flows with Gains

in the inverse graph, $f(x,y)$ flow units are delivered to vertex x and $k(x,y)f(x,y)$ flow units are removed from vertex y by arc (y,x). (Verify this for yourself using the definitions of the inverse flow.) Thus, the inverse flow has exactly the opposite effect of the original flow. Moreover, the source s in graph G becomes a sink in graph G^{-1}, and the sink t in graph G becomes a source in graph G^{-1}.

Suppose a flow $f(x,y)$ in graph G satisfies all arc capacities and supplies $b(x)$ flow units at each vertex x. Then, the corresponding flow $f^{-1}(y,x)$ in graph G^{-1} satisfies all arc capacities and supplies $-b(x)$ flow units at each vertex x. Moreover, these two flows have the same total cost.

Hence, a minimum cost flow of V units from the source in graph G^{-1} corresponds to a minimum cost flow of V units into the sink in graph G. Hence, the problem of finding a minimum cost flow with gains that delivers a given number of flow units into the sink can be solved by using the flow with gains algorithm on the inverse graph to dispatch the same number of flow units from the source of the inverse graph.

When the flow with gains algorithm is used to find a minimum cost flow that discharges V flow units from the source, there is no way to predict how many flow units ultimately arrive at the sink, since these V flow units may either arrive at the sink or be absorbed into the network. Is there any way to insure that the algorithm generates a minimum cost flow with gains in which at least W flow units arrive at the sink? Yes, simply append the original network with a new vertex T and an arc (t,T) from the original sink to vertex T. Let $l(t,T) = W$, $c(t,T) = \infty$, $k(t,T) = 1$, and $a(t,T) = 0$. Let vertex T be the sink in the appended network. Every feasible flow in the appended network corresponds to a flow in the original network that sends at least W flow units into the sink. Of course, this requirement that at least W flow units enter the sink may possibly increase the total cost of the minimum cost flow.

Suppose that no requirement is placed on how many flow units arrive at the sink. There is possibly more than one minimum cost flow that discharge V flow units from the source. Can we find the minimum cost flow that discharges V flow units from the source and simultaneously sends as many flow units as possible into the sink? This flow is generated by appending the network, as above, with a vertex T and an arc (t,T). Let $l(t,T) = 0$, $c(t,T) = \infty$, $k(t,T) = 1$, and let $a(t,T) = \varepsilon$, a very small negative number.

If ε is very, very small, the minimum cost flow generated by the flow with gains algorithm for the appended network corresponds to a minimum cost flow for the original network. If there is more than one minimum cost flow, the flow with gains algorithm when applied to the appended network will choose the minimum cost flow that sends as many flow units as possible into the sink in order to incur the small negative cost ε for each unit flowing along arc (t,T).

Conversely, to minimize the number of flow units arriving at the sink, let ε be a very, very small positive number.

For the special case when all arc traverse costs are zero, see Grinold (1973).

EXERCISES

1. List all possible flow augmenting chains from s to t in the following graph:

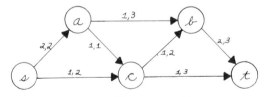

Current flow, capacity

2. Starting with the given flow, construct a maximum flow from s to t in the graph of Exercise 1. Find a minimum cut that separates s from t. Does the maximum flow saturate all arcs in this cut?

3. Describe conditions under which there exists a flow augmenting chain from the source to the sink that consists entirely of backward arcs, entirely of forward arcs.

Exercises

4. A shipping agent must decide how to maximize the total weight shipped from his warehouse to the various retail outlets he supplies. After studying Chap. 4, he is able to rephrase his problem as a maximum flow problem between a vertex representing the warehouse and the vertices representing the retail outlets.

   One of the intermediate vertices in the graph he has constructed represents a city at the junction of three highways. Due to pollution controls, this city has passed legislation limiting the tonnage that the agent can ship through the city.

   How can this additional restriction be incorporated into the graph so that the agent can use the maximum flow algorithm to solve his problem?

5. If the maximum flow algorithm discovers a flow augmenting chain that contains a backward arc, then some flow units will be removed from this backward arc and rerouted in their journey to the sink.

   (a) Is it possible for the maximum flow algorithm never to reroute any flow units? If so, what conditions generate such a situation?

   (b) Under what conditions can you be certain that the flow units assigned to a specific arc will not be rerouted during a subsequent iteration of the maximum flow algorithm.

6. Construct a minimum cost flow from s to t in the following graph using
   (a) The minimum cost flow algorithm
   (b) The out-of-kilter algorithm

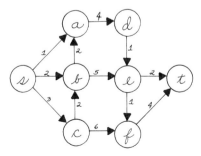

Arc costs are indicated above. All upper capacities equal one and all lower capacities equal zero.

7. Suppose that we have already calculated a minimum cost flow for a very large graph. When reviewing the final results, we notice that

   (1) An arc was omitted from the graph
   (2) The capacity of arc (x,y) was overstated by 5 units
   (3) The cost of arc (m,n) was understated by 2 units.

   (a) How can we determine which of these errors individually had an effect on the optimal solution?
   (b) Is it possible to correct individually for each of these errors without scrapping all our results? Is it possible to correct for all these errors without scrapping all our results?

8. Construct a maximum dynamic flow for 10 time periods for the graph given below. The numbers adjacent to each arc respectively denote the arc capacity and arc transit time.

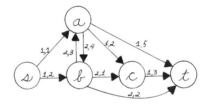

9. Suppose that the transit time of arc (a,c) in Exercise 8 changes from 2 time periods to 1 time period at the end of time period x. For which values of x, would this traverse time change leave the result of Exercise 8 unaltered?

10. Prove that a flow unit cannot travel backwards in time in a dynamic flow. Can the time expanded replica of a graph contain a circuit?

11. Generalize the maximum dynamic flow algorithm to the case when arc transit times are not necessarily integers.

12. Under what conditions will the maximum dynamic flow algorithm produce an earliest arrival flow? latest departure flow?

Exercises 177

13. Under what conditions will the number of flow units arriving at the sink during each time period remain a constant?

14. Construct a counterexample to disprove the following conjecture: Theorem 4.4 of Sect. 4.5 is valid for nonmaximum flows F' and F".

15. For the graph in Fig. 4.15, construct the four lexicographic flows described in Lemma 4.5.

16. List all absorbing and generating cycles in the following graph. All lower arc capacities equal zero, all upper arc capacities equal one. The arc gain and arc cost are indicated next to each arc.

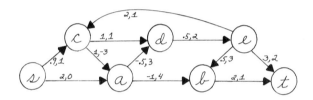

17. Use the flow with gains algorithm to find a minimum cost way to discharge 5 units from the source in the graph in Exercise 16.

18. How can the flow with gains algorithm be used to find generating cycles that can transmit flow units to the sink?

19. Prove in detail that the complementary slackness conditions remain satisfied after the vertex number change in Step 3 of the flow with gains algorithm.

20. In this chapter, we studied seven algorithms:

    (1) Flow augmenting algorithm
    (2) Maximum flow algorithm
    (3) Minimum cost flow algorithm
    (4) Out-of-kilter algorithm
    (5) Maximum dynamic flow algorithm
    (6) Earliest arrival flow algorithm
    (7) Flow with gains algorithm

    (a) Which of these algorithms can substitute for another?

(b) Which of these algorithms is a special case of another algorithm?

(c) Which of these algorithms are subroutines of another algorithm?

21. A large number of people travel by car from city A in Mexico to city B in the United States. The possible routes are shown in the following graph. The border patrol wishes to construct enough inspection stations so that every car must pass at least one inspection station. The cost of constructing an inspection station varies with location. The cost associated with each road segment is given below. What is the least cost solution to the border patrol's problem?

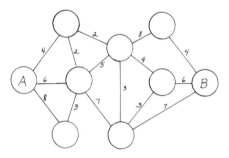

22. A management training program may be completed in a variety of different ways as shown below. Each arc represents a subprogram. Each trainee must start at s and pursue subprograms until he graduates at t. The number of trainees allowed in each subprogram is limited. Although each subprogram takes exactly one month to complete, the cost per student varies. Maximum enrollment and cost is shown in the following graph for each subprogram. What is the maximum number of students who may complete the program within the next five months so that not more than $1000 is spent on the training of any one student?

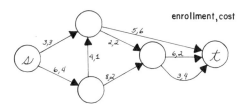

(Maximum enrollment, cost in hundreds)

REFERENCES

Edmonds, J., and R. M. Karp, 1972. Theoretical Improvements in Algorithm Efficiency for Network Flow Problems, *J. ACM,* vol. 19, no. 2, pp. 218-264.

Ford, L. R., and D. R. Fulkerson, 1962. *Flows in Networks*, Princeton Press, Princeton.

Grinold, R. C., 1973. Calculating Maximal Flows in a Network with Positive Gains, *Operations Research*, 21, pp. 528-541.

Jewell, W. S., 1962. Optimal Flow through a Network with Gains, *Operations Research*, 10, pp. 476-499.

Johnson, E. L., 1966. Networks and Basic Solutions, *Operations Research*, 14, pp. 619-623.

Maurras, J. F., 1972. Optimization of the Flow through Networks with Gains, *Math. Programming*, 3, pp. 135-144.

Minieka, E. T., 1973. Maximum, Lexicographic and Dynamic Network Flows, *Operations Research*, 21, pp. 517-527.

Minieka, E. T., 1972. Optimal Flow in a Network with Gains, *INFOR*, 10, pp. 171-178.

Truemper, K., 1973. Optimum Flow in Networks with Positive Gains, Ph.D. Thssis, Case-Western Reserve University.

Truemper, K., 1976. An Efficient Scaling Procedure for Gains Networks, *Networks*, 6, pp. 151-160.

Wilkinson, W. L., 1971. An Algorithm for Universal Maximal Dynamic Flows in a Network, *ORSA*, 19, pp. 1602-1612.

Chapter 5

MATCHING AND COVERING ALGORITHMS

5.1 INTRODUCTION

A *matching* is any set of edges in a graph such that each vertex of the graph is incdient to *at most* one edge in this set. A *covering* is any set of edges in a graph such that each vertex of the graph is incident to *at least* one edge in this set. For example, in Fig. 5.1 each of the sets {α, γ}, {β, e}, {δ, β}, {α}, {β} forms a matching. Each of the sets {α, e, γ}, {α, β, δ, e}, {β, δ, e} forms a covering. Obviously, any subset of a matching is also a matching, and any set of edges that contains a covering is also a covering.

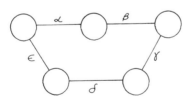

Figure 5.1

Matching and coverings have many practical applications:

EXAMPLE 1. During World War II, many airplane pilots from overrun countries fled to Britain to enlist in the Royal Air Force. Each plane sent aloft by the RAF required two pilots whose navigational skills and language skills were complementary. The RAF was interested in sending as many planes aloft at one time as possible.

Construct a graph in which each vertex represents a RAF pilot. Join together by an edge any two vertices representing pilots who can fly together. Any matching of this graph represents a possible set of planes that can simultaneously be sent aloft. Thus, the RAF was interested in finding the matching in this graph that contained the greatest possible number of edges, in other words, a *maximum cardinality matching*.

EXAMPLE 2. A real estate agency has for sale a variety of homes and a number of prospective buyers. Each prospective buyer is interested in possibly more than one of the available homes for sale. The real estate agent can estimate fairly accurately just how much each buyer would pay for each of the homes he is interested in. Since the real estate agent makes a 7% commission on each transactions he is interested in maximizing the total dollar volume of his sales. How can he accomplish this?

Let each buyer be represented by a vertex, and let each home be represented by a vertex. Join two vertices by an edge if the buyer would be willing to buy the home. Thus, each edge represents a possible transaction. Place a weight on each edge equal to the commission the real estate agent would receive for the corresponding transaction.

The real estate agent can maximize his earnings by effecting the transaction corresponding to a matching with the greatest total weight, in other words by effecting a *maximum weight matching*.

EXAMPLE 3. A committee is to be selected so that there is at least one member from each of the 50 states and at least one member from each of the 65 major ethnic groups in the United States. One hundred-seventy persons nationwide have volunteered to serve on the committee. What is the smallest committee that can be constructed from these volunteers so that all requirements are met?

Construct a graph in which each state is represented by a vertex and each ethnic group is represented by a vertex. Thus, there will

5.1 Introduction

be 50 + 65 = 115 vertices. Let each volunteer be represented by an edge joining his state to his ethnic group. Any committee that satisfies all geographic and ethnic requirements corresponds to a covering of this graph. The committee with the smallest possible membership corresponds to a covering with the smallest possible number of edges in it, in other words to a *minimum cardinality covering*.

EXAMPLE 4. A lonely hearts service permits each applicant a date with at least one other lonely heart with whom he would be compatible. Each date cost the service a different amount depending upon the specific arrangements required for it (time, place, preferences of the daters, etc.). How can the lonely hearts service meet all its obligations to its subscribers at minimum cost?

Construct a graph in which each lonely heart is represented by a vertex and each compatible couple is represented by an edge. Place a weight on each edge equal to the cost of the corresponding date. Each covering of this graph represents a way to arrange at least one acceptable date for each subscriber to the lonely hearts service. The service must find the covering with the least total cost, in other words a *minimum weight covering*.

Examples 1 to 4 have illustrated four types of useful matching and covering problems:

1. Maximum cardinality matching
2. Maximum weight matching
3. Minimum cardinality covering
4. Minimum weight covering.

Are there others?

Minimum cardinality matching. Clearly, the null matching (matching without any edges in it) is a minimum cardinality matching.

Minimum weight matching. If all edge weights are non-negative, then the null matching is a minimum weight matching. If some edge weights are negative, simply (a) delete all edges with nonnegative weight, (b) reverse the sign of the remaining edges, and (c) find a maximum weight matching on the resulting graph. The edges in this matching are clearly a minimum weight matching for the original graph. Thus, the minimum weight matching problem and the maximum weight matching problem are equivalent problems.

Maximum cardinality covering. Clearly, the entire edge set forms a maximum cardinality covering (when the graph has no isolated vertices).

Maximum weight covering. A maximum weight covering must, of course, include all edges with positive weight. If the edges with positive positive weight do not form a covering, then some edges with nonpositive weights must be included in the maximum weight covering. Which? These edges must cover the remaining uncovered vertices. These edges can be selected by finding a minimum weight covering of the remaining uncovered vertices after reversing the sign of the edge weights (i.e., making all the weights of the edges under consideration nonnegative). Thus, the minimum weight covering problem solves the maximum weight covering problem.

Happily, the maximum cardinality matching problem solves the minimum cardinality covering problem, and vice-versa. If a maximum cardinality matching M^* is known, a minimum cardinality covering C^* can be generated easily from M^*, and if a minimum cardinality covering C^* is known, a maximum cardinality matching M^* can be generated easily from C^*. How is this done?

Construction 1 (Matching to Covering): Let M be any matching. Select any exposed (unmatched) vertex v. Add to the matching any edge incident to v. Repeat this procedure until there are no more exposed vertices. The resulting set C of edges is a covering.

Construction 2 (Covering to Matching): Let C be any covering. Let v be any overcovered (i.e. incident to more than one edge) vertex in C. Remove from C any edge incident to vertex v. Repeat this procedure until there are no more overcovered vertices. The resulting set M' of edges is a matching.

THEOREM 5.1. If M is a maximum cardinality matching, then C' is a minimum cardinality covering. If C is a minimum cardinality covering, then M' is a maximum cardinality matching.

Proof: Since M is a maximum cardinality matching,

$$|C'| = |M| + (|X| - 2|M|) = |X| - |M|$$

where $|X|$ denotes the number of vertices in the graph. Since C is a minimum cardinality covering,

$$|M'| = |C| - (2|C| - |X|) = |X| - |C|$$

Thus,

$$|M| + |C'| = |X| \tag{1}$$

and

$$|C| + |M'| = |X| \tag{2}$$

Consequently, if C' is not a minimum cardinality covering, then there exists a matching generated from C' by Construction 2 that has larger cardinality than M from equation (2), which contradicts that M is a maximum cardinality matching. Also, if M' is not a maximum cardinality matching, then there exists a covering generated from M' by Construction 1 that has smaller cardinality than C from equation (1), which contradicts that C is a minimum cardinality covering. Q.E.D.

Thus, the maximum cardinality matching and minimum cardinality covering problems are equivalent. Unfortunately, no similar relationship seems to hold between the maximum weight matching problem and the minimum weight covering problem.

Section 5.2 presents an algorithm for finding a maximum cardinality matching. Section 5.3 presents an algorithm for finding a maximum weight matching. Section 5.4 presents an algorithm for finding a minimum weight covering.

5.2 MAXIMUM CARDINALITY MATCHING ALGORITHM

A *bipartite* graph is a graph whose vertex set X can be partitioned into two subsets X' and X" such that no edge in the graph joins two vertices in the same subset. Thus, each edge in a bipartite graph has one end in X' and the other end in X". For example, the graph in Fig. 5.2 is a bipartite graph with X' = {a,b,c} and X" = {d,e,f}. Note that every edge has one endpoint in X' and the other endpoint in X".

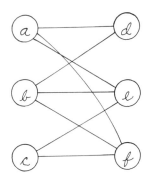

Figure 5.2
Bipartite Graph

Every cycle in a bipartite graph must contain an even number of edges (an even cycle). This follows since each edge joins vertices in different subsets and the cycle must return to the subset in which it originated.

When graph $G = (X,E)$ is a bipartite graph, the maximum cardinality matching problem can be solved easily by the maximum flow algorithm found in Sec. 4.2. This is accomplished as follows:

1. Direct all edges from X' to X"
2. Create a source vertex s and an arc (s,x) from the source to each vertex $x \in X'$
3. Create a sink vertex t and an arc (x,t) from each $x \in X''$ to the sink
4. Let each arc capacity equal 1 (see Fig. 5.3).

Call this graph G'. Since all arc capacities in G' equal 1, all flow augmenting chains will carry 1 flow unit if the maximum flow algorithm is started with a zero flow. Thus, the maximum flow algorithm will terminate with a maximum flow in which each arc carries either one flow unit or no flow units. The arcs from X' to X" in G' that carry one flow unit correspond to a matching in G. Moreover, each matching in G corresponds to a flow in G' in which the edges in the matching correspond to arcs in G' that carry one flow unit. Consequently, the matching in G corresponding to a

5.2 Maximum Cardinality Matching Algorithm

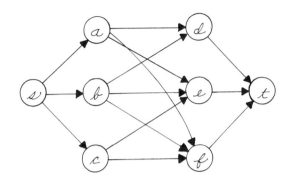

Figure 5.3
Flow Network

maximum flow in G' must be a maximum cardinality matching in G'. Otherwise, if this matching were not a maximum cardinality matching, then there would be an even larger flow in G' corresponding to the maximum cardinality matching, which is a contradiction. So, for bipartite graphs, the maximum cardinality matching problem can be solved by using the maximum flow algorithm.

If graph G is not bipartite, then G must contain a cycle with an odd number of edges in it (an odd cycle). Otherwise, the vertices of G could be partitioned into subsets X' and X" as described above. The presence of odd cycles complicates matters since now there is no obvious way to convert the matching problem into a flow problem.

How about trying to solve the maximum cardinality matching problem using linear programming? Consider the following linear programming problem:

Maximize
$$\sum_{(i,j)} x(i,j) \qquad (3)$$

such that
$$\sum_j [x(j,i) + x(i,j)] \leq 1 \quad \text{(for all vertices i)} \qquad (4)$$

$$0 \leq x(i,j) \leq 1 \quad [\text{for all edges } (i,j)] \qquad (5)$$

where $x(i,j)$ denotes the number of times edge (i,j) is used in the matching. Each constraint (4) requires that the number of matching edges incident to vertex i not exceed one.[†] Each constraint (5) requires that each edge (i,j) not be used in the matching more than once.

Clearly, every matching satisfies constraints (4) and (5) when $x(i,j) = 1$ if (i,j) is in the matching and $x(i,j) = 0$ if (i,j) is not in the matching. However, noninteger values for the variables $x(i,j)$ may also satisfy constraints (4) and (5). For example, consider the graph in Fig. 5.4. The solution

$$x(a,b) = x(b,c) = x(c,a) = 1/2$$

satisfies constraints (4) and (5). For this solution the objective function (3) equals $1/2 + 1/2 + 1/2 = 1\ 1/2$. By inspection, the

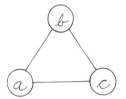

Figure 5.4

largest possible value for the objection for a solution that is a matching is 1. For example, for the matching

$$x(a,b) = 1,\ x(b,c) = x(c,a) = 0$$

the objective function (3) equals $1 + 0 + 0 = 1$. Consequently, we cannot always be sure that an optimal solution to the linear programming problem (3)-(5) will be a matching. (In Sec. 5.3, additional constraints will be added to this linear programming problem so that the optimal solution will be a matching. However, so many additional constraints will be needed that the linear programming problem would become too large to solve efficiently.)

Alas, not being able to use either flow algorithms or linear programming to solve the maximum cardinality matching problem for

---

  †Recall that an edge joining vertices i and j is denoted either by (i,j) or (j,i).

5.2 Maximum Cardinality Matching Algorithm

non-bipartite graphs, we must study an algorithm specifically designed for the maximum cardinality matching problem. The remainder of this section presents the *maximum cardinality matching algorithm* due to Edmonds (1965).

Given a matching M in graph G, an *alternating chain* is a simple chain in which the edges are alternately in and out of matching M. For example, for the matching in Fig. 5.5(a), the chain (a,b), (b,c), (c,f) is an alternating chain since the first and third edges (a,b) and (c,f) are not in M and the second edge (b,c) is in M. Also, the chain (d,a), (a,b), (b,c), (c,f) is an alternating chain since its first and third edges are in M and its second and fourth edges are not in M.

A vertex is called *exposed* if it is not incident to any matching edge. A vertex is called *matched* if it is incident to a matching edge. An *augmenting chain* is an alternating chain whose first and last vertices are exposed. For example, in Fig. 5.5(a) the chain (e,b), (b,c), (c,f) is an augmenting chain.

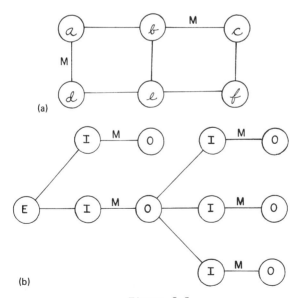

Figure 5.5

Example of (a) Augmenting Chains and (b) Alternating Tree
E = exposed, I = inner, O = outer, M = matching edge

If the edges in an augmenting chain have their roles in the matching reversed (i.e., matching edges are removed from the matching, edges not in the matching are placed in the matching), then the resulting matching contains one more edge than did the original matching. Consequently, if a matching possesses an augmenting chain, then the matching cannot be a maximum cardinality matching. Moreover, the converse of this result is also true.

THEOREM 5.2 If matching M is not a maximum cardinality matching, then matching M possesses an augmenting chain.

Proof: Let M* denote the maximum cardinality matching that has the most edges in common with M. Let G' denote the subgraph consisting of all edges that are in exactly one of these two matchings M and M*.

No vertex of G' can have more than two edges incident to it. Otherwise, two edges in the same matching would be incident to the same vertex, which is impossible. For this reason, each connected component of G' must be either a simple chain or a simple cycle.

A connected component of G' cannot be an odd cycle. Otherwise, one vertex in this odd cycle would be incident to two edges in the same matching, which is impossible. Moreover, a connected component of G' cannot be an even cycle. If a connected component of G' were an even cycle, then the role of each edge in this even cycle in M* could be reversed. This new matching would contain the same number of edges as M* but would have more edges in common with M, which is impossible. Consequently, no connected component of G' can be a cycle.

Thus, each connected component of G' is a simple chain. Consider the first and last edges in such a chain. If both of these edges are in M, then this chain is an augmenting chain in M*, which contradicts the assumption that M* is a maximum cardinality matching. If one of these edges is in M and the other is in M*, then the roles of the edges in this chain can be reversed in matching M*. This creates a new matching with the same number of edges as M* and with more edges in common with M, which is a contradiction.

5.2 Maximum Cardinality Matching Algorithm

Consequently, both the first and last edges of the chain must be in M^*, which implies that this chain is an augmenting chain in M. Q.E.D.

A matching is a maximum cardinality matching if, and only if, it does not contain an augmenting chain. The maximum cardinality algorithm is based on this result. The algorithm selects an exposed vertex and searches for an augmenting chain from this vertex to some other exposed vertex. If an augmenting chain is found, then the roles in the matching of the edges in this chain are reversed. This creates a matching with greater cardinality. If no augmenting chain is found, then another exposed vertex is similarly examined. The algorithm stops when all exposed vertices have been examined.

The maximum cardinality matching algorithm searches for an augmenting chain rooted at an exposed vertex by growing a tree rooted at this exposed vertex. This tree is called an *alternating tree* because every chain in the tree that starts at the root (exposed vertex) is an alternating chain. (The first edge in such a chain is not in the matching since the chain begins at an exposed vertex.)

The vertices in an alternating tree are called *outer* or *inner*. Consider the unique chain in the tree from the root to vertex x. If the last edge in this chain is a matching edge, then vertex x is called outer. If the last edge in this chain is not a matching edge, then vertex x is called inner. The root is called an outer vertex. See Fig. 5.5(b).

The following procedure describes how to generate an alternating tree.

Alternating Tree Subroutine

Step 1 (Initialization): Select a vertex v that is exposed in matching M. Designate vertex v as the root and label v as an outer vertex. All other vertices are unlabeled, and all edges are uncolored. (Orange edges denote the edges in the alternating tree.)

Step 2 (Growing the Tree): Select any uncolored edge (x,y) incident to any outer vertex x. (If no such edge exists, go to Step 4.) Three cases are possible:

(a) Vertex y is an inner vertex
(b) Vertex y is an outer vertex
(c) Vertex y is not labeled.

If item (a) occurs, then color (x,y) blue (i.e., not in the tree) and repeat Step 2.

If item (b) occurs, then color (x,y) orange and go to Step 3.

If item (c) occurs, color (x,y) orange. If y is an exposed vertex, stop because an augmenting chain of orange edges from v to y has been found. If y is a matched vertex, then color orange the unique matching edge (y,z) incident to y. Label vertex y as an inner vertex and label vertex z as an outer vertex. Return to Step 2.

Step 3 (Odd Cycle): This step is reached only after an edge (x,y) with two outer endpoints has been colored orange. This creates a cycle of orange edges. This cycle C must be an odd cycle since the vertices in C from x to y alternate between outer and inner vertices. Stop, because an odd cycle has been discovered.

Step 4 (Hungarian Tree): This step is reached only after no further coloring is possible. The orange edges form an alternating tree called a *Hungarian tree*. Stop.

Note that the alternating tree subroutine stops with (a) an augmenting chain, (b) an odd cycle, or (c) a Hungarian tree.

EXAMPLE 1. Let us generate an alternating tree rooted at exposed vertex a in Fig. 5.6.

Step 1: Vertex a is labeled outer.

Step 2:

Edge under examination	Outer vertices	Inner vertices	Orange edges
Initial	a	none	none
(a,h)	a,g	h	(a,h),(h,g)
(g,d)	a,g,e	h,d	(a,h),(h,g) (g,d),(d,e) |//

(continued)

5.2 Maximum Cardinality Matching Algorithm

Edge under examination	Outer vertices	Inner vertices	Orange edges
(a,b)	a,g,e,c	h,d,b	(a,h),(h,g)
			(g,d),(d,e)
			(a,b),(b,c)
(e,f)	a,g,e,c,f	h,d,b	(a,h),(h,g)
			(g,d),(d,e)
			(a,b),(b,c)
			(e,f)

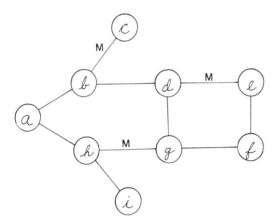

Figure 5.6
(M = matching edge)

Stop, an augmenting chain (a,h), (h,g), (g,d), (d,e), (e,f) has been found.

Note that often in Step 2 there are several edges that can be examined. The choice is arbitrary. If edge (g,f) had been examined instead of edge (g,d), the augmenting chain (a,h), (h,g), (g,f) would have been discovered. Hence, the result of the subroutine may depend upon the arbitrary choice of which edge to examine next. Lastly, note that if Step 2 colors a pair of edges orange, one of these edges is a matching edge; the other is not a matching edge.

The maximum cardinality matching algorithm can be initialized with any matching. The algorithm selects an exposed vertex and grows an alternating tree rooted at this vertex using the alternating tree subroutine. If an augmenting chain is discovered, the roles of the edges in this chain are reversed. This generates a new matching with greater cardinality and matches the exposed root of the alternating tree.

If the alternating tree turns out to be a Hungarian tree, then no augmenting chain exists from this root. (See the proof of the algorithm to verify this.) If an odd cycle is discovered, then this cycle is shrunk into a single vertex. The algorithm is continued on the new smaller graph resulting from shrinking this odd cycle.

After each exposed vertex has been examined by the alternating tree subroutine, the resulting matching is a maximum cardinality matching for the terminal graph. Next, the algorithm judiciously expands out all shrunken cycles inducing a matching on the edges of each of these cycles. After all shrunken cycles have been expanded out, the resulting matching can be shown to be a maximum cardinality matching for the original graph.

With this as motivation, we can now formally state the *maximum cardinality matching algorithm*.

Maximum Cardinality Matching Algorithm

Step 1 (Initialization): Denote the graph under consideration by G_0. Select any matching M_0 for graph G_0. All exposed vertices are called *unexamined*. Let i = 0.

Step 2 (Examination of an Exposed Vertex): If graph G_i has only one unexamined exposed vertex, to to Step 6. Otherwise, select any unexamined exposed vertex v. Use the alternating tree subroutine to grow an alternating tree rooted at vertex v.

If the alternating tree subroutine terminates with an augmenting chain, go to Step 3. If the alternating tree subroutine terminates with an odd cycle, go to Step 4. If the alternating tree subroutine terminates with a Hungarian tree, go to Step 5.

5.2 Maximum Cardinality Matching Algorithm

Step 3 (Augmenting Chain): This step is reached only after the alternating tree subroutine discovers an augmenting chain. Reverse the roles in matching M_i of the edges in the augmenting chain. This increases by one the cardinality of matching M_i and matches vertex v. Return to Step 2.

Step 4 (Odd Cycle): This step is reached only after the alternating tree subroutine discovers an odd cycle. Let $i = i + 1$. Denote the odd cycle by C_i. Shrink C_i into an artificial vertex a_i. (Recall from Chap. 2 that each edge incident to a single vertex in C_i now becomes incident to a_i.) Call the new graph G_i. Let matching M_i consist of all edges in M_{i-1} that are in G_i. (Note that all but one of the vertices in C_i are matched by edges in C_i. Hence, after C_i is shrunk into vertex a_i, at most one matching edge is incident to a_i.) Return to Step 2 selecting as the exposed vertex for examination the image of vertex v in graph G_i. Note that most of the labeling and coloring from the previous iteration of the alternating tree subroutine can be reused in the coming iteration of the alternating tree subroutine in Step 2.

Step 5 (Hungarian tree): This step is reached only after the alternating tree subroutine discovers a Hungarian tree rooted at exposed vertex v. Vertex v is now called examined. Return to Step 2.

Step 6 (Exploding Shrunken Odd Cycles): This step is reached only after all but one of the exposed vertices has been examined or matched. During the repetitions of Step 4, a sequence $G_1, G_2, \ldots, G_t$ of graphs was generated, and a sequence $a_1, a_2, \ldots, a_t$ of artificial vertices was generated. Also, a sequence $M_1, M_2, \ldots, M_t$ of matching in $G_1, G_2, \ldots, G_t$ respectively was generated.

Matching M_t is a maximum cardinality matching for graph G_t. Let $M_t^* = M_t$.

For $j = t, t-1, \ldots, 1$, generate from matching M_j^* in graph G_j a matching M_{j-1}^* in graph G_{j-1} as follows:

(a) If vertex a_j is matched in matching M_j^*, then let M_{j-1}^* consist of all edges in M_j^* together with the unique set of edges in odd cycle C_i that match all the exposed vertices in C_i (see Fig. 5.7).

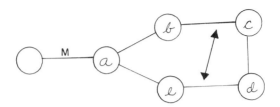

Figure 5.7

Selecting Matching Edges in Odd Cycle C_i

[Add the two edges (b,c) and (d,e) to M^*_{j-1}.]

(b) If vertex a_j is exposed in matching M^*_j, then let M^*_{j-1} consist of all edges in M^*_j together with any set of edges in C_i that match all but one of the vertices in C_i. The vertex in C_i that remains exposed is chosen arbitrarily, (see Fig. 5.8).

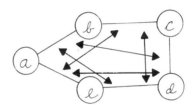

Figure 5.8

Selecting Matching Edges in Odd Cycle C_i

[Add any two nonadjacent edges to M^*_{j-1}; for example, if vertex b is to remain exposed, add the two edges (a,e) and (c,d) to the matching. All vertices in C_i are exposed in M^*_j.]

Matching M^*_0 is a maximum cardinality matching for the original graph G_0.

Proof: To prove that matching M^*_0 is a maximum cardinality matching for graph G_0 requires two parts:

1. Proof that matching M^*_t is a maximum cardinality matching for graph G_t
2. Proof that if matching M^*_i is a maximum cardinality matching for graph G_i, then matching M^*_{i-1} is a maximum cardinality matching for graph G_{i-1}.

5.2 Maximum Cardinality Matching Algorithm

Once these proofs have been accomplished, then it follows by applying item (2) successively to matchings, M_t^*, M_{t-1}^*, ..., M_0^* that M_0^* is a maximum cardinality matching for G_0.

Part (1): To show that M_t^* is a maximum cardinality matching for graph G_t, we need only show that there is no augmenting chain rooted at any exposed vertex in G_t to another exposed vertex in G_t.

Suppose that in matching M_t^* there is an augmenting chain C joining two exposed vertices v and w in G_t. At least one of these two vertices, say vertex v, must have been examined by Step 2 of the maximum cardinality matching algorithm. This examination must have terminated at Step 5 with a Hungarian tree rooted at vertex v. Traverse chain C from v to w. Let (x,y) denote the first edge in C that is not in the Hungarian tree. Thus, vertex x must be an outer vertex of the tree. Vertex y can be (a) exposed, (b) unlabeled and matched, (c) labeled outer, or (d) labeled inner.

(a) Vertex y cannot be exposed; if so, Step 2 would have terminated with an augmenting chain from v to y rather than with a Hungarian tree rooted at v.

(b) Vertex y cannot be unlabeled and matched; if so the alternating tree subroutine would have labeled vertex y as an inner vertex.

(c) Vertex y cannot be an outer vertex; if so, edge (x,y) would have been colored by the alternating tree subroutine creating an odd cycle.

(d) If vertex y is an inner vertex, then the alternating chain from root v to vertex y forms part of another augmenting chain C' from v to w. Augmenting chain C' consists of the alternating chain from v to y in the Hungarian tree together with the portion of C from y to w.

Repeat the above analysis for augmenting chain C'. This leads either to a contradiction as in cases (a), (b), and (c), or to another augmenting chain C". Each augmenting chain generated by this process must have fewer edges not in the Hungarian tree than did the preceding augmenting chain. Thus, we ultimately reach a contradiction or show that an augmenting chain has been colored by the alternating tree subroutine instead of a Hungarian tree. Consequently, if an augmenting chain exists, then the examination of vertex v cannot have terminated with a Hungarian tree, which proves Part (1).

Part (2): Suppose that matching M_{i-1}^* is not a maximum cardinality matching in graph G_{i-1}. Then, there is an augmenting chain between two exposed vertices in G_{i-1}. It is easily seen that the image of this chain in G_i is an augmenting chain with respect to matching M_i^*. This contradicts the assumption that M_i^* is a maximum cardinality matching for graph G_i, which proves Part (2). Q.E.D.

EXAMPLE 2. Let us find a maximum cardinality matching for the graph in Fig. 5.9(a). Note that this graph possesses nine vertices, and consequently, no matching can contain more than four edges.

Step 1: Start with the matching shown in Fig. 5.9(a).

Step 2: Exposed vertex a is selected for examination. Alternating tree subroutine: Label vertex a outer. Color edges (a,d) and (d,b). Label vertex d inner and label vertex b outer. Color edge (a,b) between outer vertices a and b. An odd cycle (a,d), (d,b), (b,a) has been found. Go to Step 4.

Step 4: Shrink the odd cycle into an artificial vertex a_1. The new graph G_1 is shown in Fig. 5.9(b). Return to Step 2.

Step 2: Exposed vertex a_1 (the image of a in G_1) is selected for examination. Alternating tree subroutine: Label vertex a_1 outer. Color edge (a_1,c). An augmenting chain (a_1,c) has been found. Go to Step 3.

Step 3: Add edge (a_1,c) to the matching. See Fig. 5.9(c). Return to Step 2.

Step 2: Exposed vertex g is selected for examination. Alternating tree subroutine: Label vertex g outer. Color edges (g,h) and (h,f). Label vertex h inner and label vertex f outer. Color edge (f,i). An augmenting chain (g,h), (h,f), (f,i) has been found. Go to Step 3.

5.2 Maximum Cardinality Matching Algorithm

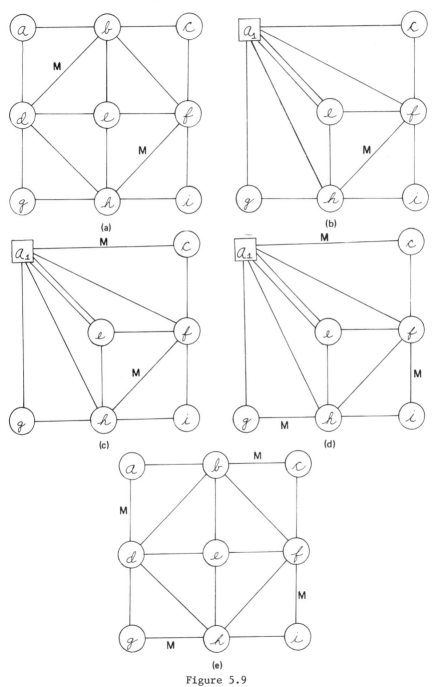

Figure 5.9

Example of the Alternating Tree Subroutine

Step 3: Remove edge (h,f) from the matching. Add edges (g,h) and (f,i) to the matching. See Fig. 5.9(d). Go to Step 2.

Step 2: Only vertex e is exposed. Go to Step 6.

Step 6: Expand vertex Q_1 back into its original odd cycle. (a,d), (d,b), (b,a). Since b is matched, add edge (a,d) to the matching. The final matching is shown in Fig. 5.9(e).

Note that neither of the edges in the original matching appear in the terminal matching. Also, note that other maximum cardinality matching are possible. For example, {(a,b), (c,f), (i,h), (d,g)} is also a maximum cardinality matching.

The terminal matching depends heavily upon the choice of which exposed vertex is examined and upon the arbitrary choices of edges colored in the alternating tree subroutine.

5.3 Maximum Weight Matching Algorithm

This section presents an algorithm due to Edmonds and Johnson (1970) for finding a maximum weight matching of a graph $G = (X,E)$. Like the maximum cardinality matching algorithm of Sec. 5.2, the basic operation of this algorithm is the generation of an alternating tree.

As we saw in Sec. 5.2, the key issue in finding a maximum cardinality matching was the existence of augmenting chains. A similar result holds for maximum weight matchings:

A *weighted augmenting chain* is an alternating chain in which:

1. The total weight of the nonmatching edges exceeds the total weight of the matching edges
2. The first vertex in the chain is exposed if the first edge in the chain is not a matching edge
3. The last vertex in the chain is exposed if the last edge in the chain is not a matching edge.

Observe that if the roles in the matching of the edges in a weighted augmenting chain are reversed, then the resulting matching has greater weight than the original matching. For example, in Fig. 5.10, chains C_1, C_2, and C_3 are alternating chains. In each chain,

5.3 Maximum Weight Matching Algorithm

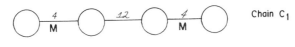

Weak Augmenting Chain

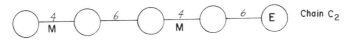

Neutral Augmenting Chain

Strong Augmenting Chain

Figure 5.10
Weighted Augmenting Chains

the nonmatching edge(s) weigh 12 units and the matching edge(s) weigh 8 units. All are weighted augmenting chains. Chain C_1 is called a *weak augmenting chain* because it contains more matching edges than nonmatching edges. Chain C_2 is called a *neutral augmenting chain* because it contains an equal number of matching and nonmatching edges. Chain C_3 is called a *strong augmenting chain* because it contains more nonmatching edges than matching edges.

Analogous to Theorem 5.2, we now show:

THEOREM 5.3 A matching M is a maximum weight matching if, and only if, M possesses no weighted augmenting chains.

Proof: If matching M possesses a weighted augmenting chain C, then M cannot be a maximum weight matching since the matching M' generated by reversing the roles of the edges in C has greater weight than matching M.

To prove the rest of the theorem, let M* be any matching with greater weight than M. As in Theorem 5.2, consider the set of all edges that are in exactly one of these two matchings M and M*. We know from the proof of Theorem 5.2 that each connected component of these edges must be either an even cycle or a chain. Since all cycles are chains, we can regard each connected component as a chain. Since M* weight more than M, in one of these chains the edges in M* must weigh more than the edges in M. This chain is a weighted augmenting chain for matching M. Q.E.D.

Let $V = \{V_1, V_2, \ldots, V_z\}$ denote the set of all odd cardinality vertex subsets. Let T_m denote the set of all edges with both endpoints in vertex set V_m. Let $T = \{T_1, T_2, \ldots, T_z\}$. Let the number of vertices in V_m be denoted by $2n_m + 1$.

No matching can contain more than n_m members of set T_m.

Let $a(i,j)$ denote the weight of edge (i,j). Let $x(i,j) = 1$ if edge (i,j) is in the matching; otherwise, let $x(i,j) = 0$.

To understand the maximum weight matching algorithm, we must first examine the following linear programming formulation of the maximum weight matching problem:

Maximize

$$\sum_{(i,j)} a(i,j) x(i,j) \tag{6}$$

$$\sum_j [x(i,j) + x(j,i)] \leq 1 \quad \text{(for all } i \in X\text{)} \tag{7}$$

$$\sum_{(i,j) \in T_m} x(i,j) \leq n_m \quad (\text{for } m = 1, 2, \ldots, z) \tag{8}$$

$$0 \leq x(i,j) \quad [\text{for all } (i,j)] \tag{9}$$

[Note that an edge joining vertices i and j is denoted either by (i,j) or (j,i).]

Constraint (7) requires that no more than one matching edge be incident to each vertex i. Constraint (8) requires that no more than n_m edges from T_m be present in the matching. The objective function (6) equals the total weight of the matching edges.

5.3 Maximum Weight Matching Algorithm

Every matching satisfies constraints (7)-(9). However, it is virtually impossible to enumerate all these constraints for graphs of even moderate size. Fortunately, the maximum weight matching algorithm produces a matching that is an optimal solution for the linear programming problem (6)-(9). How do we know that this solution is an optimal solution to the linear programming problem (6)-(9)? This is accomplished by constructing a feasible solution to the dual linear programming problem that together with the solution for the primal satisfy all complementary slackness conditions.

The dual linear programming problem for the primal linear programming problem (6)-(9) is

Minimize

$$\sum_{i \in X} y_i + \sum_{m=1}^{m=z} n_m z_m \qquad (10)$$

such that

$$y_i + y_j + \sum_{m:(i,j) \in T_m} z_m \geq a(i,j) \quad \text{[for all } (i, j)\text{]} \qquad (11)$$

$$y_i \geq 0 \quad \text{(for all } i \in X\text{)} \qquad (12)$$

$$z_m \geq 0 \quad \text{(for all } m = 1, 2, \ldots, z\text{)} \qquad (13)$$

The dual variable associated with the primal constraint (7) for vertex i is denoted by y_i. The dual variable associated with the primal constraint (8) for T_m is denoted by z_m.

The complementary slackness conditions for this pair of primal-dual linear programming problems are

$$x(i,j) > 0 \Rightarrow y_i + y_j + \sum_{m:(i,j) \in T_m} z_m = a(i, j) \quad \text{[for all } (i,j)\text{]} \qquad (14)$$

$$y_i > 0 \Rightarrow \sum_{j \in X} [x(i,j) + x(j,i)] = 1 \quad \text{(for all } i \in X\text{)} \qquad (15)$$

$$z_m > 0 \Rightarrow \sum_{(i,j) \in T_m} x(i,j) = n_m \quad \text{(for } m = 1, 2, \ldots, z\text{)} \qquad (16)$$

How does the maximum weight matching algorithm work? The algorithm starts with a null matching [all $x(i,j) = 0$] and feasible values for the dual variables y_i, $i \in X$, and z_m, $m = 1, 2, \ldots, z$, that satisfy complementary slackness conditions (14) and (16). Only conditions (15) remain unsatisfied.

At each iteration of the algorithm, the matching and/or the values of the dual variables are changed so that all the conditions (7)-(9), (11)-(13), (14), and (16) remain satisfied and so that condition (15) is satisfied for at least one more dual variable y_i. Since there are only $|X|$ dual variables y_i, after not more than $|X|$ iterations all conditions (15) become satisfied. By complementary slackness, the resulting matching must be a maximum weight matching.

Let us examine condition (15) more closely. Condition (15) states that if the dual variable y_i for vertex i is positive, then vertex i must be matched. Thus, only exposed vertices with positive dual variables violate condition (15).

The algorithm identifies an exposed vertex v with $y_v > 0$ and uses the alternating tree subroutine to grow an alternating tree rooted at vertex v. As we have seen in Sec. 5.2, the subroutine terminates with (a) an augmenting chain, (b) an odd cycle, or (c) a Hungarian tree. If the subroutine finds an augmenting chain, then this chain is a strong augmenting chain. The roles of the edges in this chain are reversed. This increases the total weight of the matching and matches vertex v. Consequently, vertex v satisfies condition (15). If the subroutine finds an odd cycle, this cycle is shrunk into an artificial vertex, and the algorithm is continued on the resulting graph. If the subroutine finds a Hungarian tree, the dual variables are changed so that all primal, dual and complementary slackness conditions, except possibly condition (15), remain satisfied and so that another edge can be added to the alternating tree. Ultimately, either y_v is reduced to zero so that condition (15) becomes satisfied or else vertex v is matched.

During the course of the algorithm, odd cycles are shrunk into artificial vertices. Eventually, all artificial vertices are expanded out into their original odd cycles. However, the vertices need not be expanded out in the same order in which they were generated. Due to these shrinkings and expansions, the algorithm will produce a sequence of graphs, G_0 G_1, ..., G_t.

5.3 Maximum Weight Matching Algorithm

With this as background, we are now prepared to state formally the *maximum weight matching algorithm*:

Maximum Weight Matching Algorithm

Step 1 (Initialization): Initially, let matching M_0 contain no edges, and let all dual variables $z_m = 0$, $m = 1, 2, \ldots, z$. Choose any initial values for the dual variables y_i, $i \in X$, such that $y_i + y_j \geq a(i,j)$ for all edges (i,j). (For instance, you could let each y_i equal half the maximum edge weight.) Let $k = 0$. Denote the original graph by $G_k = (X_k, E_k)$.

Step 2 (Examination of an Exposed Vertex): Select any non-artificial, exposed vertex v in graph G_i with $y_v > 0$. If no such vertex exists, go to Step 6. Otherwise, let E* consist of all edges (i,j) in G_k such that

$$y_i + y_j + \sum_{(i,j) \in T_m} z_m = a(i,j) \tag{17}$$

Using the alternating tree subroutine, grow an alternating tree rooted at v using only edges in E*.

If the subroutine finds an augmenting chain, go to Step 3. If the subroutine finds an odd cycle, go to Step 4. If the subroutine finds a Hungarian tree, go to Step 5.

Step 3 (Augmenting Chain): This step is reached only after the alternating tree subroutine finds an augmenting chain. Reverse the roles in matching M_k of the edges in this chain. Vertex v is no longer exposed. Return to Step 2.

Step 4 (Odd Cycle): This step is reached only after the alternating tree subroutine finds an odd cycle. Let $k = k + 1$. Denote this odd cycle by C_k. Shrink the odd cycle C_k into an artificial vertex a_k. Denote the new graph by $G_k = (X_k, E_k)$. Let M_k be the matching in G_k consisting of all edges in M_{k-1} that are in G_k.

In all future labeling, let all vertices subsumed into artificial vertex a_k carry the same label as a_k.

Return to Step 2 and continue to grow an alternating tree rooted at the image of vertex v in G_k even if this vertex is artificial. Note that the labeling and coloring of the last iteration of the alternating tree subroutine can be salvaged for the next iteration.

Step 5 (Hungarian Tree): This step is reached only after the alternating tree subroutine finds a Hungarian tree.
Let

$$d_1 = \min\{y_i + y_j - a(i,j)\} \qquad (18)$$

where the minimization is taken over all (i,j), where $i \in X_0$ is an outer vertex and $j \in X_0$ is unlabeled.
Let

$$d_2 = 1/2 \min\{y_i + y_j - a(i,j)\} \qquad (19)$$

where the minimization is taken over all (i,j), where $i \in X_0$ is an outer vertex and $j \in X_0$ is an outer vertex, and i and j are not inside the same artificial vertex.
Let

$$d_3 = 1/2 \min\{z_m\} \qquad (20)$$

where the minimization is taken over all odd cardinality vertex sets V_m that are shrunk into an artificial vertex a_k that is labeled inner.
Let

$$d_4 = \min\{y_i\} \qquad (21)$$

where the minimization is taken over all vertices $i \in X_0$ that are labeled outer.
Lastly, let

$$d = \min\{d_1, d_2, d_3, d_4\} \qquad (22)$$

Adjust the dual variables as follows:

(a) Outer vertex variables y_i are decreased by d
(b) Inner vertex variables y_i are increased by d

5.3 Maximum Weight Matching Algorithm

(c) For each outer artificial vertex in G_k, increase its dual variable z_m by $2d$.

(d) For each inner artificial vertex in G_k, decrease its dual variable z_m by $2d$.

If $d = d_1$, then the edge (i,j) that determined d_1 enters E^*. This edge can now be colored by the alternating tree subroutine. Return to Step 2 and continue to grow an alternating tree rooted at v.

If $d = d_2$, then the edge (i,j) that determined d_2 enters E^*. This edge can now be colored by the alternating tree subroutine creating an odd cycle. Return to Step 2 and continue to grow an alternating tree rooted at v.

If $d = d_3$, then some dual variable z_i becomes zero. Expand the artificial vertex corresponding to this dual variable back to its original odd cycle. Let $k = k + 1$. Call the resulting graph $G_k = (X_k, E_k)$. Let matching M_k consist of all edges in M_{k-1} together with the n_i edges of T_i that match the $2n_i$ exposed vertices of V_i. (The remaining vertex of V_i is matched in M_k since all inner artificial vertices in G_{k-1} are matched in M_{k-1}.)

Return to Step 2 and continue to grow an alternating tree rooted at v.

If $d = d_4$, then the dual variable y_i of some outer vertex i becomes zero. The chain in the alternating tree froom root v to vertex i is a neutral augmenting chain. Reverse the roles in matching M_k of the edges in this chain. Vertex v becomes matched and vertex i becomes exposed, which is all right since $y_i = 0$. Return to Step 2.

Step 6 (Expansion of Artificial Vertices): This step is reached only after all vertices violating condition (15) have been examined by Step 2. Consider all artificial vertices remaining in the terminal graph. Expand out each artificial vertex in reverse order (the last to be generated is expanded first, etc.) and induce a maximum matching on the resulting odd cycle.

The terminal matching is a maximum weight matching for the original graph G_0. Stop.

Proof of the Maximum Weight Matching Algorithm: The algorithm starts with a matching with no edges and maintains a matching throughout all iterations. We need only prove that the terminal matching is a maximum weight matching. This is accomplished by showing that the terminal values for the dual variables y_i, $i \in X$, and z_m, $m = 1, 2, \ldots, z$, satisfy dual feasibility (11)-(13) and complementary slackness (14)-(16).

Since equations (18)-(22) assure that no dual variable is ever reduced to a negative value, and since Step 1 selects initial dual variable values that are nonnegative, conditions (12) and (13) are satisfied at all times by the algorithm.

To verify that condition (11) is satisfied at all times by the algorithm, note that

1. Initially, condition (11) is satisfied
2. Edge (i,j) must be (a) in an artificial vertex, (b) not in an artificial vertex and in E*, or (c) not in an artificial vertex and not in E*.

If (a) occurs, then a dual variable change does not change the left side of condition (11) since both y_i and y_j change by d and the dual variable for the artificial vertex containing (i,j) changes by $-2d$.

If (b) occurs, then a dual variable change increases the inner vertex dual variable and decreases the outer vertex dual variable. Hence, the left side of (11) remains unchanged

If (c) occurs, then

$$y_i + y_j + \sum_{m:(i,j) \in T_m} z_m > a(i,j)$$

If both i and j are not labeled, or if i and j have different labels, then the dual variable change preserves the above inequality. If one of the vertices i,j is labeled inner and the other is unlabeled, or if both vertices i and j are labeled inner, then the left side of (11) increases after a dual variable change. If one of the vertices i and j is labeled outer and the other is unlabeled, then by equation

5.3 Maximum Weight Matching Algorithm

(18) the dual variable will not reverse the above inequality. If both vertices i and j are outer vertices, then by equation (19) the dual variable change will not reverse the preceding inequality.

Thus, under all dual variable changes condition (11) remains satisfied.

To verify condition (14), we must consider two cases: (a) matching edge (i,j) is not in an artificial vertex at the beginning of Step 6, or (b) matching edge (i,j) is contained in an artificial vertex at the beginning of Step 6.

If (a) occurs, then edge (i,j) ∈ E* and condition (14) is satisfied. If (b) occurs, then edge (i,j) is contained in some artificial vertex after the last dual variable change has been made, and condition (11) holds with equality. Hence, condition (14) is satisfied by the terminal dual variable values.

Condition (15) is satisfied for all vertices at the end of the algorithm; otherwise, Step 2 would have been repeated for any vertex violating condition (15).

Lastly, it remains to show that condition (16) is satisfied. The only way that a dual variable z_m can become positive is to shrink the vertices in V_m into an artificial vertex. In Step 6, each artificial vertex is expanded and a maximum matching is induced on the corresponding odd cycle. Thus, condition (16) is satisfied. Q.E.D.

EXAMPLE 3. Let us find a maximum weight matching for the graph in Fig. 5.11.

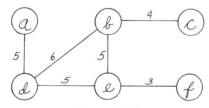

Figure 5.11
Maximum Weight Matching Algorithm

Step 1: Since the largest edge weight equals 6, let $y_i = 6/2 = 3$ for all vertices i. Let all $z_i = 0$. Initially, no edges are in the matching.

Step 2: Exposed vertex d is selected for examination. Alternating tree subroutine: $E^* = \{(d,b)\}$. Vertex d is labeled outer. Vertex b is labeled, and an augmenting chain (d,b) has been discovered. Go to Step 3.

Step 3: Add edge (d,b) to the matching.

Step 2: Exposed vertex e is selected for examination. Alternating tree subroutine: $E^* = \{(d,b)\}$. Vertex e is labeled outer. No further labeling is possible. The alternating tree consisting entirely of vertex e is Hungarian. Go to Step 5.

Step 5: Perform a dual variable change.

$$d_1 = m\{y_d + y_e - a(d,e), y_b + y_e - a(b,e), y_f - y_e - a(f,e)\}$$
$$= \min\{3 + 3 - 5, 3 + 3 - 3\} = 1$$
$$d_2 = \infty, d_3 = \infty$$
$$d_4 = y_e = 3$$
$$d = \min\{d_1, d_2, d_3, d_4\} = \min\{1, \infty, \infty, 3\} = 1$$

Thus, y_e decreases by 1. (See Fig. 5.12.)

	y_a	y_b	y_c	y_d	y_e	y_f	z_g	Matching
Initialization	3	3	3	3	3	3	0	Empty
Examine d	3	3	3	3	3	3	0	(d,b)
Examine e	3	3	3	3	2	3	0	(d,b)
Examine g	3	2	3	2	1	3	2	(a,g)
Examine c	3	2	2	2	1	3	2	(a,g)
	2	3	1	3	2	3	0	(a,d), (e,b)
Examine f	2	3	1	3	2	1	0	(a,d), (f,e)
								(b,c)

Figure 5.12

Dual Variable Changes

Maximum Weight Matching Algorithm

5.3 Maximum Weight Matching Algorithm

Step 2: Continued examination of exposed vertex e. Alternating tree subroutine: $E^* = \{(d,b), (d,e), (e,b)\}$. Vertex e is labeled outer. Vertex b is labeled inner, vertex d is labeled outer, and edges (e,b) and (d,b) are colored. Next, edge (d,e) joining outer vertices d and e is colored. An odd cycle has been discovered. Go to Step 4.

Step 4: Shrink the odd cycle (e,b), (d,b), (d,e) into an artificial vertex g. Denote the dual variable for the odd vertex set e, b, d by z_g. The new graph resulting from this shrinking is shown in Fig. 5.13.

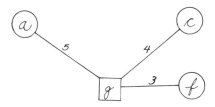

Figure 5.13
Graph Resulting from Shrinking an Odd Cycle

Step 2: Examination of exposed artificial vertex g. Alternating tree subroutine: $E^* = \{(d,b), (d,e), (e,b)\}$. Vertex g is labeled outer. (Hence, vertices d, b, e are also labeled outer.) No further labeling is possible. The tree consisting entirely of vertex g is Hungarian. Go to Step 5.

Step 5: Perform a dual variable change.

$d_1 = \min\{y_b + y_c - a(b,c), y_a + y_d - a(a,d), y_e + y_f - a(e,f)\}$
    $= \min\{3 + 3 - 4, 3 + 3 - 5, 2 + 3 - 3\} = \min\{2,1,2\} = 1$

$d_2 = \infty$

$d_3 = \infty$ (there are no inner artificial vertices)

$d_4 = \min\{y_b, y_d, y_e\} = \min\{3,3,2\} = 2$

$d = \min\{1, \infty, \infty, 2\} = 1$

Thus, $y_d = 3 - 1 = 2$, $y_e = 2 - 1 = 1$, $y_b = 3 - 1 = 2$, $z_g = 2d = 2$.

Note that for each edge in E^*, condition (11) is satisfied with equality, e.g., for edge (d,e),

$$y_d + y_e + z_g = 2 + 1 + 2 = 5 = a(d,e)$$

Step 2: Continued examination of exposed artificial vertex g. Alternating tree subroutine: $E^* = \{(d,e), (e,b), (b,d), (a,d)\}$. Vertex g is labeled outer. Thus, vertices b, e, d are also labeled outer. Edge (a,g) is colored, and an augmenting chain has been found. Go to Step 3.

Step 3: Add edge (a,g) to the matching.

Step 2: Examination of exposed vertex c. Alternating tree subroutine: $E^* = \{(d,e), (e,b), (b,d), (a,d)\}$. Vertex c is labeled outer. No further labeling. The tree consisting entirely of vertex c is Hungarian. Go to Step 5.

Step 5: Perform a dual variable change.

$$d_1 = y_c + y_b - a(c,d) = 3 + 2 - 4 = 1$$

$$d_2 = \infty$$

$$d_3 = \infty$$

$$d_4 = y_c = 3$$

$$d = \min\{1, \infty, \infty, 3\} = 1$$

Thus, $y_c = 3 - 1 = 2$.

Step 2: Continued examination of exposed vertex c. Alternating tree subroutine: $E^* = \{(d,e), (e,b), (b,d), (a,d), (c,b)\}$. Vertex c is labeled outer. Vertex g is labeled inner, and vertex a is labeled outer. Edges (c,g) and (g,a) are colored. (Since g is an inner vertex, vertices b, e, d are also labeled inner.) No further labeling is possible. The tree is Hungarian. Go to Step 5.

Step 5: Perform a dual variable change.

$$d_1 = \infty$$

$$d_2 = \infty$$

5.3 Maximum Weight Matching Algorithm

$$d_3 = (1/2)z_g = 1$$
$$d_4 = \min\{y_c, y_a\} = \min\{2, 3\} = 2$$
$$d = \min\{\infty, \infty, 1, 2\} = 1$$

Thus, $y_c = 2 - 1 = 1$, $y_a = 3 - 1 = 2$, $y_b = 2 + 1 = 3$, $y_d = 2 + 1 = 3$, $y_e = 1 + 1 = 2$, and $z_g = 2 - 2(1) = 0$.

Since z_g has returned to zero, artificial vertex g must be expanded. Expanding vertex g yields the original graph in Fig. 5.11.

Now, we must induce a maximum matching on the odd cycle corresponding to vertex g. Since edge (d,a) is a matching edge, only vertices e and b are left exposed. Thus, edge (e,b) is added to the matching.

Step 2: Examination of exposed vertex f. Alternating tree subroutine: $E^* = \{(d,e), (e,b), (b,d), (a,d), (b,c)\}$. Vertex f is labeled outer. No further labeling is possible. The tree consisting entirely of vertex f is Hungarian. Go to Step 5.

Step 5: Perform a dual variable change.

$$d_1 = y_f + y_e - a(f,e) = 3 + 2 - 3 = 2$$
$$d_2 = \infty$$
$$d_3 = \infty$$
$$d_4 = y_f = 3$$
$$d = \min\{2, \infty, \infty, 3\} = 2.$$

Thus, $y_f = 3 - 2 = 1$. All other dual variables remain unchanged.

Step 2: Continued examination of exposed vertex f. Alternating tree subroutine: E^* consists of all edges in graph G. Vertex f is labeled outer. Vertex e is labeled inner, and vertex b is labeled outer. Edges (f,e) and (e,b) are colored. Edge (b,c) is colored next. An augmenting chain (f,e), (e,b), (b,c) has been found. Go to Step 3.

Step 3: Reverse the roles of the edges in chain (f,e), (e,b), (b,c). Thus, edge (e,b) leaves the matching, and edges (f,e) and (b,c) enter the matching. The matching now consists of edges (a,d), (f,e), and (b,c).

Step 2: There are no more unexamined, exposed vertices. Go to Step 6.

Step 6: No artificial vertices remain in the final graph. Stop.

The maximum weight matching is (a,d), (f,e), (b,c) with a total weight $5 + 3 + 4 = 12$.

Notice that at termination all dual variables $z_m = 0$, and $\sum y_i = 2 + 3 + 1 + 3 + 2 + 1 = 12$. Thus, the value of the primal objective function (6) is 12, and the value of the dual objective function (10) is also 12. Consequently, by equation (4) of Chap. 1, both primal and dual solution must be optimal.

5.4 Minimum Weight Covering Algorithm

This section presents an algorithm (White, 1967) that finds a minimum weight covering of graph $G = (X,E)$. Like the matching algorithms in Secs. 5.2 and 5.3, the basic operation of this algorithm is the growing of an alternating tree using the alternating tree subroutine.

As before, let $V = \{V_1, V_2, \ldots, V_z\}$ denote the set of all odd cardinality subsets of X. Let $2n_m + 1$ denote the number of vertices in subset V_m. Let U_m denote the set of all edges with *one or both* endpoints in X_m.

Every covering must contain at least n_m+1 edges in U_m. As before; let $a(i,j)$ denote the weight of edge (i,j). Let $x(i,j) = 1$ if edge (i,j) is in the covering; otherwise, let $x(i,j) = 0$.

For the present, assume that all edge weights are positive. Consider the following linear programming problem:

Minimize
$$\sum_{(i,j)} a(i,j)x(i,j) \tag{23}$$
such that
$$\sum_{j} [x(i,j) + x(j,i)] \geq 1 \quad \text{(for all } i \in X\text{)} \tag{24}$$

5.4 Minimum Weight Covering Algorithm

$$\sum_{(i,j) \in U_m} x(i,j) \geq n_m + 1 \quad \text{(for } m = 1, 2, \ldots, z\text{)} \tag{25}$$

$$x(i,j) \geq 0 \quad [\text{for all } (i,j)] \tag{26}$$

Constraint (24) requires that at least one edge in the covering be incident to vertex i. Constraint (25) requires that at least $n_m + 1$ edges from set U_m be present in the covering. Expression (23) equals the total weight of the covering.

Clearly, any covering satisfies constraints (24)-(26). The dual linear programming problem to (23)-(26) is:

Maximize

$$\sum_{i \in X} y_i + \sum_{m=1}^{m=z} (n_m + 1) z_m \tag{27}$$

such that

$$y_i + y_j + \sum_{m:(i,j) \in U_m} z_m \geq a(i,j) \quad [\text{for all } (i,j)] \tag{28}$$

$$y_i \geq 0 \quad \text{(for all } i \in X\text{)} \tag{29}$$

$$z_m \geq 0 \quad \text{(for } m = 1, 2, \ldots, z\text{)} \tag{30}$$

The complementary slackness conditions for the primal-dual pair of linear programming problems are

$$x(i,j) > 0 \Rightarrow y_i + y_j + \sum_{m:(i,j) \in U_m} z_m = a(i,j) \quad [\text{for all } (i,j)] \tag{31}$$

$$y_i > 0 \Rightarrow \sum_{j \in X} [x(i,j) + x(j,i)] = 1 \quad \text{(for all } i \in X\text{)}$$

that is, vertex i is covered only once,

$$z_m > 0 \Rightarrow \sum_{(i,j) \in U_m} x(i,j) = n_m + 1 \quad \text{(for } m = 1, 2, \ldots, z\text{)} \tag{33}$$

That is, the $2n_m + 1$ vertices in X_m are covered by n_m edges with both endpoints in X_m and one edge with only one endpoint in X_m.

Note that the dual variable y_i corresponds to constraint (24) for vertex i, and hence, y_i can be regarded as the dual variable for vertex i. Also, note that the dual variable z_m corresponds to constraint (25) for odd vertex subset V_m and can be regarded as the dual variable for vertex subset V_m. Thus, there is a dual variable associated with each vertex and with each odd cardinality subset of vertices.

The minimum weight covering algorithm will shrink odd cycles into artificial vertices as did the previous algorithms in this chapter. Thus, each artificial vertex a_k will contain an odd cardinality subset X_k of vertices. The dual variable z_k associated with X_k can be regarded as the dual variable for artificial vertex a_k. From constraints (28)-(30),

$$y_i \leq \min_j \{a(i,j), a(j,i)\}$$

Vertex y is called *saturated* if equality holds, i.e., if y_i is as large as possible. Otherwise, vertex i is called *unsaturated*. If $y_i = 0$, then vertex i is called *empty*. Note that condition (32) does not apply to empty vertices, and hence, only empty vertices can have more than one edge in the covering incident to them. Also, note that each saturated vertex must be joined by an edge to an empty vertex.

The minimum weight covering algorithm consists of two phases: Phase 1 generates a matching; Phase 2 converts this matching into a covering, which is a minimum weight covering.

Phase 1 is initialized with all primal variables $x(i,j) = 0$, i.e., a null matching, with all dual variables $z_m = 0$, and with any values for the dual variables y_i that satisfy (28) and (29). Each iteration of Phase 1 examines an exposed, unsaturated vertex v. Vertex v becomes matched, or the dual variables are altered so that vertex v becomes saturated. Conditions (28)-(31) and (33) remain satisfied throughout. The algorithm moves to Phase 2 after all exposed, unsaturated vertices have been eliminated.

Phase 2 converts the terminal matching generated by Phase 1 into a covering by adding to the matching each edge joining an exposed saturated vertex to its empty vertex. The resulting covering satisfies all primal, dual and complementary slackness conditions, and hence, this covering is a minimum weight covering.

5.4 Minimum Weight Covering Algorithm

How does Phase 1 alter the matching and/or dual variables so that vertex v is no longer exposed or no longer unsaturated? This is accomplished by using the alternating tree subroutine to grow an alternating tree[†] rooted at vertex v. As before, the subroutine terminates with (a) an augmenting chain, (b) an odd cycle, or (c) a Hungarian tree. If (a) occurs, the roles of the edges in the matching are reversed along this augmenting chain, and vertex v becomes matched. If (b) occurs, then the odd cycle is shrunk into an artificial vertex, and the algorithm is continued on the shrunken replica of the original graph. If (c) occurs, the dual variables are altered so that (c1) a new edge can be added to the alternating tree, or (c2) an outer vertex j becomes saturated. If (c1) occurs, the subroutine continues to grow an alternating tree rooted at v. If (c2) occurs, the alternating chain from v to j forms a neutral augmenting chain. The roles in the matching of the edges in this chain are reversed. Vertex v becomes matched, and vertex j becomes an exposed, saturated vertex.

Ultimately, vertex v is either matched or saturated.

The algorithm shrinks odd cycles into artificial vertices. Eventually, the algorithm expands back all artificial vertices into their original odd cycles. However, the artificial vertices are expanded back in an order only somewhat related to the order in which they were formed. Consequently, the algorithm works on a sequence $G_0, G_1, \ldots, G_t$ of graphs.

With this as background, we are now ready to state formally the *minimum weight covering algorithm*.

Minimum Weight Covering Algorithm

Phase 1 (Matching).

Step 1 (Initialization): Let $k = 0$. Let $G_k = (X_k, E_k)$ denote the graph under consideration. Let matching M_k contain no edges.

[†] The alternating tree uses only edges that satisfy (28) with equality.

To save writing, define

$$w_i = y_i + \sum_{i \in V_n} z_n \quad \text{(for all } i \in X_0\text{)} \tag{34}$$

Let all dual variables $z_m = 0$, $m = 1, 2, \ldots, t$. Choose any values for the dual variables y_i such that

$$w_i + w_j \leq a(i,j) \quad [\text{for all } (i,j) \in E_0] \tag{35}$$

$$w_i \geq 0 \quad \text{(for all } i \in X_0\text{)} \tag{36}$$

For example, $y_i = 0$ for all i will meet all requirements.

Step 2 (Examination of an Exposed, Unsaturated Vertex):

Let $E^* = \{(i,j): w_i + w_j = a(i,j), i \in X_0, j \in X_0\}$.

Select any exposed, unsaturated vertex $v \in X_k$ that is not an artificial vertex. If no such vertex exists, go to Phase 2. Otherwise, use the alternating tree subroutine to grow an alternating tree rooted at v using only the edges of E^* present in graph G_k.

If the alternating tree subroutine terminates with an augmenting chain, go to Step 3. If the alternating tree subroutine terminates with an odd cycle, go to Step 4. If the alternating tree subroutine terminates with a Hungarian tree, go to Step 5.

Step 3 (Augmenting Chain): This step is reached only after the alternating tree subroutine finds an augmenting chain. Reverse the roles in matching M_k of the edges in this chain. The exposed root vertex v is no longer exposed. Return to Step 2 to examine another exposed, unsaturated vertex.

Step 4 (Odd Cycle): This step is reached only after the alternating tree subroutine finds an odd cycle. Let $k = k + 1$. Shrink this odd cycle into an artificial vertex a_k. Denote the new graph resulting from this shrinking by G_k. Let M_k be the matching consisting of all edges in matching M_{k-1} that appear in graph G_k.

In all future labelings in the alternating tree subroutine, let all vertices subsumed into artificial vertex a_k carry the same label as a_k.

5.4 Minimum Weight Covering Algorithm

Return to Step 2 and continue to grow an alternating tree rooted at the image of vertex v in G_k^*†. In this case, the labeling and coloring of the last iteration of the alternating tree subroutine can be salvaged for the next iteration of this subroutine.

Step 5 (Hungarian Tree): This step is reached only after the alternating tree subroutine terminates with a Hungarian tree. Let

$$d_1 = \min\{a(i,j) - w_i - w_j\} \qquad (37)$$

where the minimization is taken over all outer vertices i in X_0 and all unlabeled vertices j in X_0.

Let

$$d_2 = 1/2 \min\{a(i,j) - w_i - w_j\} \qquad (38)$$

where the minimization is taken over all outer vertices $i \in X_0$ and outer vertices $j \in X_0$ such that $(i,j) \notin E^*$. Let

$$d_3 = 1/2 \min\{z_m\} \qquad (39)$$

where the minimization is taken over all m, where the vertices V_m constitute an inner artificial vertex. Let

$$d_4 = \min\{a(i,j) - w_i\} \qquad (40)$$

where the minimization is taken over all nonartificial outer vertices $i \in X_k$ and all inner vertices $j \in X_0$, where $(i,j) \in E^*$. Let

$$d_5 = \min\{w_i - \sum_{m \in V_m} z_m\} \qquad (41)$$

where vertex $i \in X_0$ is contained in an outer artificial vertex.

Lastly, let

$$d = \min\{d_1, d_2, d_3, d_4, d_5\}. \qquad (42)$$

Adjust the following dual variable as follows:

1. Outer vertex variables w_i are increased by d.
2. Inner vertex variables w_i are decreased by d.

---

†This is the only time when an artificial vertex is the root of the alternating tree.

3. Increase the dual variable z_m corresponding to each outer artificial vertex in G_k by 2d.
4. Decrease the dual variable z_m corresponding to each inner artificial vertex in G_k by 2d.

If $d = d_1$, then an additional edge is added to E* and the alternating tree can be grown further. Return to Step 2 to continue growing an alternating tree rooted at v.

If $d = d_2$, then an additional edge is added to E*. This edge forms an odd cycle in E*. Return to Step 2 to continue growing an alternating tree rooted at v.

If $d = d_3$, then the dual variable z_i for some inner artificial vertex returns to zero value. Let k = k + 1. Expand this artificial vertex back into its original odd cycle. Since the artificial vertex was an inner vertex, one matching edge is incident to it. Add edges in the odd cycle to the matching so that the remaining vertices in the odd cycle are all matched. Call the new graph G_k and the new matching M_k. Return to Step 2 to continue growing an alternating tree rooted at v.

If $d = d_4$, then some outer vertex j becomes saturated. The alternating chain in the alternating tree from root v to vertex j is a neutral augmenting chain. Reverse the roles in matching M_k of the edges in this chain from v to j. Vertex j becomes an exposed, saturated vertex. Vertex v is no longer exposed. Return to Step 2 to examine another exposed, unsaturated vertex.

If $d = d_5$, then y_i becomes zero for some vertex i contained in an outer artificial vertex a_m. The alternating chain from root vertex v to a_m forms a neutral augmenting chain. Reverse the roles in the matching of the edges in this chain from v to a_m. Vertex v becomes matched, and artificial vertex a_m becomes exposed. Return to Step 2 to examine another exposed, unsaturated vertex.

Phase 2 (Converting the Matching into a Covering).

Step 7 (Exposed Nonartificial Vertices): This step is reached only after Step 2 has eliminated all exposed, unsaturated nonartificial vertices in graph G_k. Add to M_k the edge joining each exposed,

5.4 Minimum Weight Covering Algorithm

saturated, nonartificial vertex in G_k to its empty vertex. (The empty vertex is not contained in any matched artificial vertex. See Lemma 5.2.) Now, the only exposed vertices in G_k are artificial vertices. Go to Step 8.

Step 8 (Expansion of Artificial Vertices): If graph G_k contains no artificial vertices, stop because M_k is a minimum weight covering for graph G_0. Otherwise, expand the last remaining artificial vertex a_m to be formed back into its original odd cycle. Let $k = k + 1$. Call the new graph resulting from this expansion G_k.

Two cases are possible: (a) vertex a_m was covered at the end of Phase 1, or (b) vertex a_m was exposed at the end of Phase 1.

If case (a) occurs, then one vertex in the odd cycle corresponding to a_m is matched, and the other vertices in the odd cycle corresponding to a_m are exposed. Let M_k consist of all the edges in M_{k-1} together with edges from the odd cycle corresponding to a_m that match the exposed vertices in this cycle. Repeat Step 8.

If case (b) occurs, then $y_i = 0$, for some vertex i in the odd cycle corresponding to a_m. (See Lemma 5.1.) Let M_k consist of all edges in M_{k-1} together with both edges in the odd cycle incident to vertex i and a matching of the remaining vertices in the odd cycle. Repeat Step 8.

Before proving the algorithm, two lemmas are needed.

LEMMA 5.1 At the end of Phase 1, each exposed artificial vertex contains an empty vertex.

Proof: Let a_m denote any exposed artificial vertex present at the end of Phase 1. When a_m was first generated, it was either exposed or matched. If a_m was exposed, then it was the root of the current alternating tree. This alternating tree was not discarded until after a_m was matched or until a_m contained an empty vertex. See Step 5 for $d = d_5$.

If a_m was matched, the only time that a_m could become exposed was when a_m contained an empty vertex. See Step 5 for $d = d_5$.

Lastly, if a_m is shrunk into another artificial vertex a_n, then a_n must have been expanded out before the end of Phase 1. After a_n is expanded out, a_m is again matched. See Step 5 for $d = d_3$. Q.E.D.

LEMMA 5.2 At the end of Phase 1, the empty vertex corresponding to a saturated, exposed vertex is not contained in a matched artificial vertex.

Proof: At the end of Phase 1, suppose that vertex i is a saturated, exposed vertex and vertex j is its empty vertex. We must show that vertex j is not contained in any matched artificial vertex. Since $y_i = a(i,j)$, $y_j = 0$, and $\sum_{m:i \in V_m} z_m = 0$, it follows from from (28) and (34) that $\sum_{m:j \in V_m} z_m = 0$. Thus, the dual variable for any artificial vertex a_m that contains vertex j must equal zero.

Consider the last time that artificial vertex a_m was labeled. If a_m was labeled inner, then a_m would have been expanded out since Step 5 expands out all inner artificial vertices with dual varaible equal to zero. If a_m was labeled outer, then the chain from the root to a_m was a neutral augmenting chain and a_m would have been left exposed. See Step 5 for $d = d_5$. Either way, artificial vertex a_m cannot be matched. Q.E.D.

Proof of the Minimum Weight Covering Algorithm: Phase 1 generates a matching; Phase 2 converts the matching to a covering for graph G_0. We need only show that this terminal covering is a minimum weight covering. This is accomplished by showing that the terminal values for the dual variables y_i and z_m together with the $x(i,j)$ values for the terminal covering satisfy dual feasibility (28)-(30) and complementary slackness (31)-(33).

Proof for (28): Condition (28) can be rewritten as

$$y_i + y_j + \sum_{(i,j) \in U_k} z_k = y_i + y_j + \sum_{m:i \in V_m} z_m + \sum_{m:j \in V_m} z_m - \sum_{m:i,j \in V_m} z_m$$

$$= w_i + w_j - \sum_{m:i,j \in V_m} z_m \leq a(i,j) \quad (28')$$

5.4 Minimum Weight Covering Algorithm

Initially, all $z_m = 0$ and $w_i + w_j \leq a(i,j)$. Condition (28') is satisfied with strict inequality until edge (i,j) becomes a member of set E*. After edge (i,j) joins E*, vertices i and j have (a) different labels, (b) are in the same artificial vertex, or (c) have no labels.

If (a) occurs, then at each dual variable change, the change in the y_i is offset by the opposite change in y_j, and $z_m = 0$ for all m for which $i \in V_m$ and $j \in V_m$. If (b) occurs, then w_i changes by d, w_j changes by d and the dual variable z of the artificial vertex containing i and j has an offsetting change of 2d. If (c) occurs, then w_i, w_j and all z_m, $i \in V_m$, $j \in V_m$ remain unchanged. Hence, condition (28') always remains satisfied.

Proof of (29): From the definition of d_5 in equation (41), the algorithm always maintains

$$w_i \geq \sum_{m: i \in V_m} z_m$$

Hence, $y_i \geq 0$ at all iterations of the algorithm.

Proof of (30): From the definition if d_3 in equation (39), z_k never is reduced below zero.

Proof of (31): Edge (i,j) is placed into the final covering in (a) Phase 1, (b) Step 7, or (c) Step 8.

If (a) occurs, then edge (i,j) is in E* and vertices i and j are not in the same artificial vertex when Phase 1 ends. Thus, $\sum_{m: i, j \in V_m} z_m = 0$, and from (28'), it follows that condition (31) is satisfied.

If (b) occurs, then vertex i is saturated and vertex j is empty, and both vertices are nonartificial at the end of Phase 1. Hence, $y_i = a(i,j)$, and $y_j = 0$, and condition (31) is satisfied.

If (c) occurs, then edge (i,j) is contained in an artificial vertex at the end of Phase 1. Hence, at the end of Phase 1, edge $(i,j) \in$ E*, and condition (31) is satisfied.

Proof for (32): At the end of Phase 1, all vertices are matched or saturated. In Step 7, the only vertices that are covered a second time are empty vertices. In Step 8, the only vertices that are covered a second time are empty vertices. Hence, only empty vertices are covered more than once, and all other vertices are covered exactly once.

Proof for (33): If $z_m > 0$ at the end of phase 1, then the vertices in V_m are shrunk into an artificial vertex a_t at the end of Phase 1. The vertices in V_t are covered in Step 8 by n_t edges from the odd cycle plus an additional edge. Thus, $n_t + 1$ edges of U_m are present in the final covering and condition (33) follows. Q.E.D.

EXAMPLE 4. Let us find a minimum weight covering for the graph in Fig. 5.14. The weight of each edge appears next to the edge.

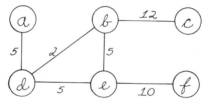

Figure 5.14

Minimum Weight Covering Algorithm

Step 1 (Initialization): Let all $z_m = 0$, $m = 1, 2, \ldots, z$. Let all $w_i = 1/2 \min\{a(i,j)\} = 1$. Initially, no edges are in the matching.

Step 2 (Examination of Exposed, Unsaturated Vertex d): Alternating tree subroutine: $E^* = \{(d,b)\}$. Vertex d is labeled outer. Edge (d,b) is colored. An augmenting chain has been found.

Step 3: Reverse the roles of the edges in the augmenting chain (d,b). Edge (d,b) is added to the matching.

Step 2 (Examination of Exposed, Unsaturated Vertex e): Alternating tree subroutine: $E^* = \{(d,b)\}$. Vertex e is labeled outer. No further labeling is possible. The tree is Hungarian. Go to Step 5.

5.4 Minimum Weight Covering Algorithm

Step 5: Perform a dual variable change.

$$d_1 = \min\{a(e,b) - w_e - w_b,\ a(e,d) - w_e - w_d,\ a(e,f) - w_e - w_f\}$$
$$= \min\{5 - 1 - 1,\ 5 - 1 - 1,\ 10 - 1 - 1\} = 3$$
$$d_2 = d_3 = d_4 = d_5 = \infty$$
$$d = \min\{d_1, d_2, d_3, d_4, d_5\} = 3$$

Now, $w_e = 1 + 3 = 4$. No other dual variables change value. (See Fig. 5.15.)

	w_a	w_b	w_c	w_d	w_e	w_f	z_1	Edges
Initialization	1	1	1	1	1	1	0	Empty
Examine d	1	1	1	1	1	1	0	(d,b)
Examine e	1	1	1	1	4	1	0	Empty
Examine a_1	1	2	1	2	5	1	2	Empty
Examine a	3	2	1	2	5	1	2	(a, a_1)
Examine c	3	2	10	2	5	1	2	(a, a_1)
Examine c again	4	1	11	1	4	1	0	(a,d), (b,e)
Examine c again	5	0	12	0	5	1	0	(a,d), (b,e)
Examine f	5	0	12	0	5	5	0	(a,d), (b,c)
								(e,f)
Phase 2	5	0	12	0	5	5	0	(a,d), (b,c)
								(e,f)

Figure 5.15

Dual Variable Changes

Minimum Weight Covering Algorithm

Step 2: (Continued Examination of Vertex e): Alternating tree subroutine: $E^* = \{(d,b), (e,b), (d,e)\}$. Vertex e is labeled outer. Edges (e,b) and (b,d) are colored, vertex b is labeled inner, and vertex d is labeled outer. Edge (d,e) is colored. An odd cycle (e,b), (b,d), (d,e) has been found. Go to Step 4.

Step 4: Shrink the odd cycle (e,b), (b,d), (d,e) into an artificial vertex a_1. See Fig. 5.16. Let $z_1 = 0$ denote the dual variable for a_1. Now, the matching is empty.

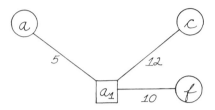

Figure 5.16

Graph Resulting from Shrinking an Odd Cycle

Step 2: Examination of exposed artificial vertex a_1. Alternating tree subroutine: $E^* = \{(e,b), (b,d), (d,e)\}$. Vertex a_1 is labeled outer, and consequently, vertices e, b, d must also be labeled outer. No further labeling is possible. The tree consisting entirely of vertex a_1 is Hungarian. Go to Step 5.

Step 5: Perform a dual variable change.

$$d_1 = \min\{a(e,f) - w_e - w_f,\ a(b,c) - w_b - w_c,\ a(a,d) - w_a - w_d\}$$
$$= \min\{10 - 1 - 1,\ 12 - 1 - 1,\ 5 - 1 - 1\} = 3$$
$$d_2 = d_3 = d_4 = \infty$$
$$d_5 = \min\{w_e,\ w_b,\ w_d\} = \min\{4,\ 1,\ 4\} = 1$$
$$d = \min\{d_1, d_2, d_3, d_4, d_5\} = 1$$

The dual variables become

$$w_e = 4 + 1 = 5$$
$$w_b = 1 + 1 = 2$$
$$w_d = 1 + 1 = 2$$
$$z_1 = 0 + 2(1) = 2$$

Since $d = d_5$, vertex a_1 can be left exposed. Note that $y_b = w_b - z_1 = 2 - 2 = 0$, and $y_d = w_d - z_1 = 2 - 2 = 0$.

Step 2: Examination of exposed, unsaturated vertex a. Alternating tree subroutine: $E^* = \{(e,b), (b,d), (d,e)\}$. Vertex a is labeled outer. No further labeling is possible. The tree consisting entirely of vertex a is a Hungarian tree. Go to Step 5.

5.4 Minimum Weight Covering Algorithm

Step 5: Perform a dual variable change.

$d_1 = \min\{a(a,d) - w_a - w_d\} = 5 - 1 - 2 = 2$

$d_2 = d_3 = d_4 = d_5 = \infty$

$d = d_1 = 2$

Dual variable $w_a = 1 + 2 = 3$. Return to Step 2.

Step 2: Continued examination of exposed, unsaturated vertex a. Alternating tree subroutine: E* = {(d,b), (e,b), (d,e), (a,d)}. Vertex a is labeled outer. Vertex a_1 is colored. The edge (a, a_1) is an augmenting chain. Add this edge to the matching.

Step 2: Examination of exposed, unsaturated vertex c. Alternating tree subroutine: E* is as in preceding Step 2. Vertex c is labeled outer. No further labeling is possible. The tree consisting entirely of vertex c is a Hungarian tree. Go to Step 5.

Step 5: Perform a dual variable change.

$d_1 = \min\{a(c,b) - w_c - w_b\} = 12 - 1 - 2 = 9$

$d_2 = d_3 = d_4 = d_5 = \infty$

$d = d_1 = 9$

Dual variable $w_c = 1 + 9 = 10$. Return to Step 2.

Step 2: Continued examination of exposed, unsaturated vertex c. Alternating tree subroutine. E* = {(d,b), (e,b), (d,e), (a,d), (c,b)}. Vertex c is labeled outer. Vertex a_1 is labeled inner, vertex a is labeled outer, and edges (c,b) and (a_1,a) are colored. No further labeling is possible. The tree consisting of (c,b), (a_1,a) is Hungarian. Go to Step 5.

Step 5: Perform a dual variable change.

$d_1 = d_2 = d_5 = \infty$

$d_3 = 1/2 \min\{z_1\} = 1/2(2) = 1$

$d_4 = \min\{a(c,b) - w_c, a(a,d) - w_a\} = \min\{12 - 10, 5 - 3\} = 2$

$d = d_3 = 1$

The dual variables become

$w_a = 3 + 1 = 4$

$w_b = 2 - 1 = 1$

$w_c = 10 + 1 = 11$

$w_d = 2 - 1 = 1$

$w_e = 5 - 1 = 4$

$w_f = 1$

$z_1 = 2 - 2(1) = 0$

Since $d = d_3$, z_1 becomes zero, and inner artificial vertex a_1 can be expanded back to its original odd cycle. The graph resulting from the expansion of a_1 is the original graph of Fig. 5.14. Let the matching consist of the previous matching edge (a,d) together with edge (b,e) that matches the remaining two exposed vertices d and e of the odd cycle (b,d), (d,e), (e,b). Return to Step 2.

Step 2: Continued examination of exposed, unsaturated vertex c. Alternating tree subroutine: E* consists of all edges in the graph except (e,f). Vertex c is labeled outer. Vertex b is labeled inner, vertex e is labeled outer, and edges (c,b) and (b,e) are colored. Next, vertex d is labeled inner, vertex a is labeled outer, and edges (e,d) and (a,d) are colored. No further labeling is possible. The current tree consisting of edges (c,d), (b,e), (e,d), (d,a) is Hungarian. Go to Step 5.

Step 5: Perform a dual variable change.

$d_1 = \min\{a(e,f) - w_e - w_f\} = 10 - 4 - 1 = 5$

$d_2 = d_3 = d_5 = \infty$

$d_4 = \min\{a(c,b) - w_c, a(e,b) - w_e, a(e,d) - w_e, a(a,d) - w_a\}$

$ = \min\{12 - 11, 5 - 4, 5 - 4, 5 - 4\} = 1$

$d = d_4 = 1$

The dual variables become

5.4 Minimum Weight Covering Algorithm

$w_a = 4 + 1 = 5$

$w_b = 1 - 1 = 0$

$w_c = 11 + 1 = 12$

$w_d = 1 - 1 = 0$

$w_e = 4 + 1 = 5$

$w_f = 1$

$z_1 = 0$

Vertex c is now saturated and need not be examined further.

Step 2: Examination of exposed, unsaturated vertex f. Alternating tree subroutine: $E^* = \{(a,d), (d,e), (e,b), (b,c)\}$. Vertex f is labeled outer. No further labeling is possible. The tree consisting entirely of vertex f is Hungarian. Go to Step 5.

Step 5: Perform a dual variable change.

$d_1 = a(f,e) - w_f - w_e = 10 - 1 - 5 = 4$

$d_2 = d_3 = d_4 = d_5 = \infty$

$d = d_1 = 4$

Dual variable $w_f = 1 + 4 = 5$.

Step 2: Continued examination of exposed, unsaturated vertex f. Alternating tree subroutine: $E^* = \{(a,d), (d,e), (e,b), (b,c), (e,f)\}$. Vertex f is labeled outer. Next, vertex e is labeled inner and vertex b is labeled outer, and edges (f,e), and (e,b) are colored. Next vertex c is labeled and edge (b,c) is colored. An augmenting chain (f,e), (e,b), (b,c) has been discovered. Go to Step 3.

Step 3: Reverse the roles in the matching of the edges in the augmenting chain. Edges (f,e) and (b,c) are added to the matching, and edge (e,b) is removed from the matching. Now, the matching consists of edges (a,d), (f,e), (b,c).

Step 2: There are no more exposed vertices. Go to Phase 2.

Phase 2: There are no more artificial vertices. Stop, the current set of edges is a minimum weight covering.

The weight of the covering consisting of edges (a,d), (f,e), (b,c) is 5 + 10 + 12. The value of the dual objective function (27) at termination is $\sum y_i = \sum w_i = 5 + 0 + 12 + 0 + 5 + 5 = 27$. Thus, the terminal objective function values of the primal and dual linear programming problems are both 27, and the by equation (4) in Chapter 1, the primal and dual solutions must be optimal for their respective problems.

Negative Edge Weights

The minimum weight covering algorithm assumed that all edge weights are nonnegative. [Otherwise, the linear programming problem (23)-(26) from which the algorithm was derived would achieve an optimal solution whenever $x(i,j) = \infty$ when $a(i,j) < 0$.]

Obviously, every edge with negative weight must be present in a minimum weight covering. Moreover, every edge with zero weight may also be present in a minimum weight covering.

How can we find a minimum weight covering when edges with negative weight are present in the graph? Simply change each negative edge weight to zero. Next, perform the minimum weight covering algorithm using these new nonnegative edge weights. Add to the terminal covering all edges not present in the terminal solution that originally had negative weight. The resulting covering is a minimum weight covering.

Note that if $a(i,j) = 0$, then at all times $w_i = 0$, $w_j = 0$, $w_i + w_j = 0 = a(i,j)$ and hence $(i,j) \in E^*$ at all times. Moreover, vertices i and j are always saturated and need never be examined by Step 2 of the algorithm.

EXERCISES

1. Construct a graph in which a maximum cardinality matching is not a maximum weight matching. Construct a graph in which a minimum cardinality covering is not a minimum weight covering.
2. For the bipartite graph in Fig. 5.17, construct

    (a) A maximum cardinality matching
    (b) A maximum weight matching
    (c) A minimum cardinality covering
    (d) A minimum weight covering

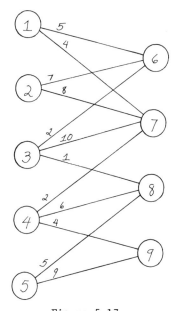

Figure 5.17

3. For the graph in Fig. 5.18, construct

    (a) A maximum cardinality matching
    (b) A maximum weight matching
    (c) A minimum cardinality covering
    (d) A minimum weight covering

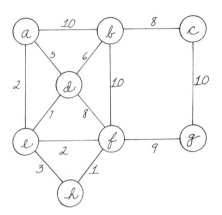

Figure 5.18

4. How does the maximum cardinality matching algorithm remove an edge that has incorrectly been placed into solution? How does the maximum weight matching algorithm accomplish this? How does the minimum weight covering algorithm accomplish this?

5. A tour group of thirty persons has just arrived at a hotel. Each room in the hotel has two twin beds. The hotel manager wishes to assign the guests to as few rooms as possible without placing two unrelated persons of the opposite sex in the same room. Combinations of roommates such as husband-wife, father-daughter, brother-sister are acceptable. How can the hotel manager solve his problem?

6. What are the chief similarities and differences between the maximum weight matching algorithm and Phase 1 of the minimum weight covering algorithm?

7. A machine shop possesses six different drilling machines. On a certain day, five jobs that need drilling arrived. The number of man-hours required to perform each job on each of the machines is given in the following. Find the best way to assign each job to a different machine.

		A	B	Job C	D	E
Machine	1	5	7	6	4	9
	2	8	10	3	4	7
	3	6	11	5	4	7
	4	5	8	7	3	9
	5	3	6	4	2	7
	6	3	7	5	3	7

8. The United Nations sponsors a sister-cities program in which cities are paired off for cultural and educational exchange programs. This year 10 new cities have applied. What is the best way to pair them off so that the total distance between all sister cities is minimized? The intercity distances are

		1	2	3	4	5	6	7	8	9	10
From city	1	0	80	70	70	60	45	90	110	85	155
	2		0	75	95	90	80	90	160	70	45
	3			0	65	70	60	100	80	80	55
	4				0	80	80	70	170	200	250
	5					0	110	170	190	270	300
	6						0	100	150	110	200
	7							0	75	95	100
	8								0	90	100
	9									0	50

REFERENCES

Balinski, M., 1969. Labelling to Obtain a Maximum Matching, *Combinatorial Mathematics and Its Applications* (Bose and Dowling, eds.), University of North Carolina Press, Chapel Hill, pp. 585-602.

Berge, C., 1957. Two Theorems in Graph Theory, *Proc. Natl. Acad. Sci. U.S.A.,* Vol. 43, pp. 842-844.

Brown, J. R., Maximum Cardinality Matching, Kent State University, unpublished manuscript.

Edmonds, J., 1965. Paths, Trees, and Flowers, *Can. J. Math.*, Vol. 17, pp. 449-467.

Edmonds, J., 1965. Maximum Matching and Polyhedra with 0-1 Vertices, *J. Res. N.B.S.*, vol. 69B, no. 1, 2, pp. 125-130.

Edmonds, J., and Ellis Johnson, 1970. Matching: A Well Solved Class of Integer Linear Programs, *Combinatorial Structures and Their Applications*, Gordon and Breach, New York, pp. 89-92.

White, L. J., 1967. A Parametric Study of Matchings and Coverings in Weighted Graphs, Ph.D. Thesis, University of Michigan.

Chapter 6

POSTMAN PROBLEM

6.1 INTRODUCTION

Before starting his route, a postman must pick up his letters at the post office, then he must deliver letters along each block in his route, and finally he must return to the post office to return all undelivered letters. Wishing to conserve energy, every postman would like to cover his route with as little walking as possible. In nongraph terms, the postman problem is the problem of how to cover all the streets in the route and return back to the starting point with as little traveling as possible. Obviously, not only postmen, but many kinds of carriers encounter such a problem. For example, a policeman wants to know the most efficient way to patrol all the streets on his beat, a farmer wants to know the best route for seeding his fields, and a track repair crew needs to know the best way to cover all the tracks. The first work on this problem appeared in a Chinese journal which called the problem the postman problem. Sometimes, it is referred to as the Chinese postman problem.

The postman problem can be restated in graph terms. Construct a graph $G = (X,E)$ in which each edge represents a street in the postman's route and each vertex represents a junction between two streets. The postman problem is the problem of finding the shortest route for the postman so that he traverses each edge at least once and returns to his starting vertex.

Let s denote the starting vertex, and let a(i,j) > 0 denote the length of edge (i,j).

There are several ways that the postman can traverse all the edges in the graph in Fig. 6.1 and return back to vertex s. For example, each of the following four routes will do this:

Route 1:   (s,a), (a,b), (b,c), (c,d), (d,b), (b,s)
Route 2:   (s,a), (a,b), (b,d), (d,c), (c,b), (b,s)
Route 3:   (s,b), (b,c), (c,d), (d,b), (b,a), (a,s)
Route 4:   (s,b), (b,d), (d,c), (c,b), (b,a), (a,s)

Each of these four routes traverses each edge exactly once; thus, the total length of each route is 3 + 2 + 1 + 3 + 7 + 6 = 22. The postman cannot do any better than this.

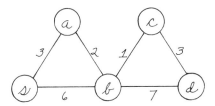

Figure 6.1
Graph with an Euler Tour

A route in which each edge is traversed exactly once is called an *Euler tour* after the mathematician Leonhard Euler, mentioned in Chap. 1 in connection with the Konigsberg bridge problem.

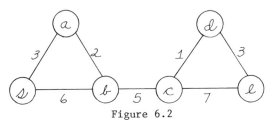

Figure 6.2
Graph with No Euler Tour

6.1 Introduction

Consider the graph in Fig. 6.2. Obviously, there is no way for the postman to traverse edge (b,c) only once. No euler tour is possible for this graph. An optimal (i.e., shortest total length) route for this graph is (s,a), (a,b), (b,c), (c,d), (d,e), (e,c), (c,b), (b,s). The total length of this tour is $3 + 2 + 5 + 1 + 3 + 7 + 5 + 6 = 32$. Are there any other optimal routes?

Obviously, no postman route, let alone an Euler tour, exists when the graph is disconnected (i.e., has several components). Consequently, for the remainder of this chapter, we shall always assume that the graph under consideration is a connected graph.

The number of times that a postman arrives at a vertex must equal the number of times that the postman departs from that vertex. If the postman does not repeat any edges incident to a vertex, then this vertex must have an even number of edges incident to it. Let the total number of edges incident to vertex x be called the *degree* of vertex x and be denoted by $d(x)$. If all vertices in graph G have even degree, then graph G is called *even*.

In a directed graph, let the number of arcs directed into vertex x be called the *inner degree* of vertex x and be denoted by $d^-(x)$. Let the number of arcs directed away from vertex x be called the *outer degree* of vertex x and be denoted by $d^+(x)$. In Fig. 6.8, $d^-(a) = 2$, $d^+(a) = 1$. If $d^+(x) = d^-(x)$ for all vertices x in graph G, then graph G is called *symmetric*.

Suppose we know an optimal postman route in graph G that starts and ends at vertex s. How can we find an optimal postman route for a different starting vertex, say vertex t? This is accomplished as follows: Any optimal route R that starts at vertex s eventually encounters vertex t for the first time. Call this part of the route R_1. Call the remainder of the route R_2. Note that R_1 starts at s and ends at t, and R_2 starts at t and ends at s. Form a new route R' consisting of R_2 followed by R_1. Route R' starts at t and ends at t and has the same total length as R. Consequently, R' must be an optimal route starting from vertex t. Thus, we can conclude:

THEOREM 6.1 The total length of an optimal postman route is the same for every starting vertex.

Section 2 describes how to find an optimal postman route when the graph is undirected (the edges are like two-way streets since they can be traversed in either direction). Section 3 describes how to find an optimal postman route when the graph is directed (the edges are like one-way streets since they can only be traversed in a specified direction). Section 4 considers the postman problem in a graph in which some arcs are directed and others are undirected (both one-way streets and two-way streets).

6.2 Postman Problem for Undirected Graphs

This section describes how to solve the postman problem for any undirected graph $G = (X,E)$, in which the edges can be traversed in either direction.

Two cases must be considered separately:

Case A: Graph G is even.
Case B: Graph G is not even.

Case A: If graph G is even, then an optimal solution to the postman problem is an Euler tour. The postman does not have to repeat any edge.

How can we find an Euler tour of graph G that starts at the starting vertex s? Traverse any edge (s,x) incident to vertex s. Next, traverse any unused edge incident to vertex s. Repeat this process of traversing unused edges until you return to vertex s. (The process must return to vertex s since every vertex has even degree and every visit to a vertex leaves an even number of unused edges incident to that vertex. Hence, every time a vertex is entered, there is an unused edge for departing from that vertex.) The traversed edges constitute a cycle C_1. If all edges were used in cycle C_1, then stop because C_1 is an Euler tour of graph G. Otherwise, generate another cycle C_2 of unused edges starting with any unused edge. Continue generating cycles C_3, C_4, ..., from the unused edges until all edges have been used.

6.2 Postman Problem for Undirected Graphs

Next, splice together all these cycles C_1, C_2, ..., into one cycle C that contains all the edges in G. Cycle C contains each edge exactly once and is an optimal solution to the postman problem.

More specifically, two cycles C_1 and C_2 can be spliced together to form one cycle only if they share a common vertex x. This is accomplished as follows: Start at any edge of cycle C_1 and travel along C_1 until vertex x is encountered. Then detour and traverse all the edges of C_2 returning to vertex x. Lastly, continue traveling the edges of C_1 until you arrive back at the initial edge. The route traveled is the cycle formed by splicing C_1 and C_2 together. This procedure can be extended in the obvious way to splice together any number of cycles into one cycle as long as the cycles cannot be split into two subsets that have no vertices in common.

EXAMPLE 1. Let us find an optimal postman route for the even graph in Fig. 6.3. Starting at vertex s, let us travel the perimeter of the graph until we return to vertex s. This generates the cycle

$$C_1 = (s,b), (b,c), (c,f), (f,i), (i,h), (h,g), (g,d), (d,s)$$

Consider the edges in C_1 as used edges. Next, starting with unused edge (d,b), let us travel the unused upper triangular cycle C_2 = (d,b), (b,e), (e,d). Consider the edges in C_2 as used edges. Next, starting with the unused edge (e,f) let us travel the lower triangular cycle C_3 = (e,f), (f,h), (h,e). Now all edges have been used

Splice together these cycles as follows:

Insert C_2 between (g,d) and (d,s) in C_1.
Insert C_3 between (b,e) and (e,d) in C_2.

The result is C = (s,b), (b,c), (c,f), (f,i), (i,h), (h,g), (g.d), (d,b), (b,e), (e,f), (f,h), (h,e), (e,d), (d,s). Clearly, C is an optimal solution to the postman problem for this graph since C contains each edge exactly once and begins and ends at vertex s. Note that the edge lengths were not considered.

Case B: Graph G is not even. In any postman route, the number of times that the postman enters a vertex equals the number of times

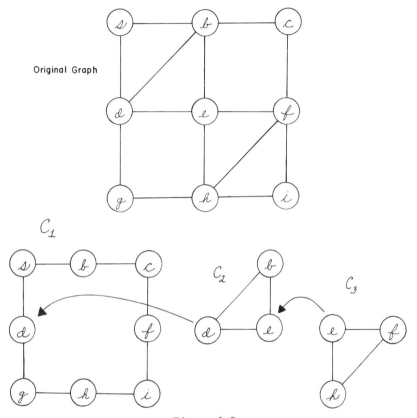

Figure 6.3

Splicing Cycles Together

that the postman leaves that vertex. Consequently, if vertex x does not have even degree, then at least one edge incident to vertex x must be repeated by the postman.

Let f(i,j) denote the number of times that edge (i,j) is repeated by the postman. Edge (i,j) is traversed f(i,j) + 1 times by the postman. Of course, f(i,j) must be a nonnegative integer Note that f(i,j) contains no information about the direction of travel across edge (i,j).

Construct a new graph G* = (X,E*) that contains f(i,j) + 1 copies of each edge (i,j) in graph G. Clearly, an Euler tour of graph G* corresponds to a postman route in graph G.

6.2 Postman Problem for Undirected Graphs

The postman wishes to select values for the $f(i,j)$ variables so that
(a) Graph G* is an even graph
(b) $\Sigma\, a(i,j)f(i,j)$, the total length of repeated edges, is minimized.

If vertex x is an odd degree vertex in graph G, then an odd number of edges incident to vertex x must be repeated by the postman, so that in graph G* vertex x has even degree. Similarly, if vertex x is an even degree vertex in graph G, than an even number of edges (zero is an even number) incident to vertex x must be repeated by the postman, so that in graph G* vertex x has even degree. Recall from Chap. 1, Exercise 2, that graph G contains an even number of vertices with odd degree.

If we trace out as far as possible a chain of repeated edges starting from an odd-degree vertex, this chain must necessarily end at another odd-degree vertex. Thus, the repeated edges form chains whose initial and terminal vertices are odd-degree vertices. Of course, any such chain may contain an even-degree vertex as one of its intermediate vertices. Consequently, the postman must decide (a) which odd-degree vertices will be joined together by a chain of repeated edges, and (b) the precise composition of each such chain.

By performing either the Floyd or Dantzig shortest path algorithm of Sec. 3.2, the postman can determine a shortest chain between each pair of odd-degree vertices in graph G.

The postman can determine which pairs of odd-degree vertices are to be joined by a chain of repeated edges as follows: Construct a graph $G' = (X',E')$ whose vertex set consists of all odd-degree vertices in G and whose edge set contains an edge joining each pair of vertices. Let the weight of each edge equal a very large number minus the length of a shortest path between the corresponding two vertices in graph G as found by the Floyd or Dantzig algorithm.

Next, the postman should find a maximum weight matching for graph G' using the maximum weight matching algorithm found in Sec. 5.4. Since graph G' has an even number of vertices and each pair of vertices in G' is joined by an edge, the maximum weight matching will cover each vertex exactly once. This matching matches together odd-degree vertices in graph G. The edges in a shortest chain joining a matched pair of odd-degree vertices should be repeated by the postman. Since this matching has maximum total weight, the resulting postman route must have minimum total length.

Thus, we can solve the postman problem for an undirected graph by using the Floyd or Dantzig algorithm and the maximum weight matching algorithm. No new algorithm is needed.

EXAMPLE 2. Let us find an optimal postman route for the undirected graph in Fig. 6.4. Notice that vertices a, c, d, and f have odd degree. The length of a shortest chain between all pairs of odd vertices is shown in Fig. 6.5. The reader should verify these values using either the Floyd or Dantzig algorithm.

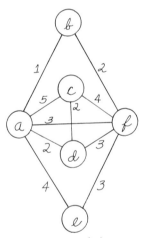

Figure 6.4

Form the graph G' shown in Fig. 6.6. The vertices of G' are the odd-degree vertices a, c, d, and f of graph G. All possible edges are present in G'. Happily, since G' does not have many vertices, we can find a minimum weight matching of all the edges in G' by enumeration rather than by using the maximum weight matching algorithm. Three matchings are possible:

6.2 Postman Problem for Undirected Graphs

Matching	Weight
(a,c), (d,f)	4 + 3 = 7
(a,d), (c,f)	2 + 4 = 6
(a,f), (c,d)	3 + 2 = 5

	a	b	c	d	e	f
a	0	1	4	2	4	3
b	1	0	5	3	5	2
c	4	5	0	2	7	4
d	2	3	2	0	6	3
e	4	5	7	6	0	3
f	3	2	4	3	3	0

Figure 6.5

Shortest Path Length Matrix

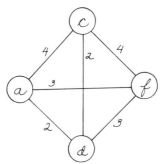

Figure 6.6

Graph G'

Consequently, the minimum weight matching is (a,f), (c,d). Thus, the postman should repeat the shortest path from a to f, which is edge (a,f) and should repeat the shortest path from c to d, which is edge (c,d). Figure 6.7 shows graph G* in which edges (a,f) and (c,d) have each been duplicated once. All vertices in G* have even degree, and an optimal postman route for the original graph in Fig. 6.4 corresponds to an Euler tour of graph G* in Fig. 6.7. The technique described in Case A for even graphs can be applied to graph G*. An optimal route is (a,b), (b,f), (f,e), (e,a), (a,c), (c,f),

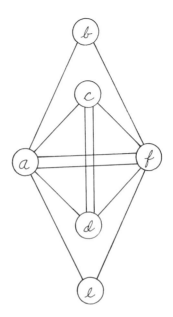

Figure 6.7
Graph G*

(f,d), (d,c), (c,d), (d,a), (a,f), (f,a) which traverses each edge in G* exactly once and traverses each edge in G at least once. Only edge (a,f) and (c,d) are repeated in graph G. The total length of this route is 34 units, which is 5 units more than the sum of the edge lengths.

Note that if the maximum weight matching algorithm had been used, then each edge weight in graph G' would have been set equal to a large number, say M, minus the length of a shortest path between the corresponding two endpoints in G. Thus, the matching (a,c), (d,f) would have a total weight equal to $(M - 4) + (M - 3) = 2M - 7$. Matching (a,d), (c,f) would have total weight equal to $(M - 2) + (M - 4) = 2M - 6$. Matching (a,f), (c,d) would have total weight equal to $(M - 3) + (M - 2) = 2M - 5$, and would be selected as the maximum weight matching.

The large M values can be viewed simply as a device for converting the problem of finding a minimum weight matching that covers all vertices into a maximum weight matching problem.

6.3 POSTMAN PROBLEM FOR DIRECTED GRAPHS

In this section we shall study the postman problem for a directed graph G = (X,A). A directed graph corresponds to a physical situation in which all streets are one-way streets. The direction of an arc specifies the direction in which the corresponding street must be traversed.

Unlike the postman problem for undirected graphs, the postman problem may have no solution for a directed graph. For example, consider the graph in Fig. 6.8. Once the postman arrives at either vertex a or vertex b he cannot return to vertex s because no arcs leave set {a,b} for a vertex not in this set. In general, no solution exists for the postman problem whenever there is a set S of vertices with the property that no arcs go from a vertex in S to a vertex not in S. If no such set S exists, then it is always possible for the postman to complete his route, no matter how long it takes, since he cannot be trapped anywhere.

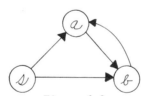

Figure 6.8

No Postman Route Exists

As before, the number of times that a postman enters a vertex must equal the number of times that the postman leaves that vertex. Consequently, if vertex x has more arcs entering it than leaving it [that is, $d^-(x) > d^+(x)$], then the postman must repeat some of the arcs leaving x. Similarly, if vertex x has more arcs leaving it than entering it [that is, $d^+(x) > d^-(x)$], the postman must repeat some arcs entering x. Thus, if for some vertex x, $d^+(x) \neq d^-(x)$, then no Euler tour is possible.

Two cases must be considered separately:

Case A: Graph G is symmetric [that is, $d^+(x) = d^-(x)$ for all x].
Case B: Graph G is not symmetric.

Case A: If graph G is symmetric, then it is possible for the postman to perform his route without repeating any arcs, i.e., the optimal solution to the postman problem is an Euler tour.

An Euler tour of graph $G = (X,A)$ can be found by using a technique similar to the technique for finding an Euler tour of an even undirected graph. Starting at the starting vertex s, traverse the arcs along their direction without reusing any arc until you return to vertex s. This traces out a circuit C_1. Next, starting at any unused arc, trace out another circuit C_2 using only unused arcs. Repeat this procedure until all arcs have been used. Lastly, splice together all the circuits into one large circuit C as done in Fig. 6.3. Circuit C contains each arc exactly once and constitutes an optimal solution to the postman problem for Case A.

Case B: As before, let f(i,j) denote the number of times that the postman repeats arc (i,j). The postman wants to select non-negative integer values for the f(i,j) variables so as to minimize

$$\sum a(i,j)\, f(i,j) \tag{1}$$

the total length of repeated arcs such that he enters and leaves each vertex x the same number of times, that is,

$$d^-(i) + \sum_j f(j,i) = d^+(i) + \sum_j f(i,j) \tag{2}$$

Rewriting equation (2) yields

$$\sum_j [f(i,j) - f(j,i)] = d^-(i) - d^+(i) \equiv \underline{D}(i) \tag{3}$$

Thus, the postman wishes to minimize expression (1) such that equation (3) is satisfied for all vertices i in graph G. This minimization problem is merely a minimum cost flow problem in disguise. The vertices with $\underline{D}(i) < 0$ are sinks with demand equal to $-\underline{D}(i)$. The vertices with $\underline{D}(i) > 0$ are sources with supply equal to $\underline{D}(i)$. The

6.3 Postman Problem for Directed Graphs

vertices with $\underline{D}(i) = 0$ are intermediate vertices. All arc capacities are infinite.

This minimum cost flow problem can be solved by appending a supersource and a supersink to the graph, connecting the supersource to all sources, connecting the supersink to all sinks. Let the capacity of each arc leaving the supersource equal the supply at its terminal vertex. Let the capacity of each arc into the supersink equal the demand of its initial vertex. Then solve for a minimum cost flow that satisfies all source and sink requirements by using the minimum cost flow algorithm of Sec. 4.3.

Since all right-side values in equations (3) are integers, we know that the minimum cost flow algorithm will produce optimal values for the $f(i,j)$ that are nonnegative integers.

After finding the optimal integer values for the $f(i,j)$ variables, create a graph G* with $f(i,j) + 1$ copies of arc (i,j) for all $(i,j) \in A$. By equation (3), graph G* is symmetric. The technique described for Case A can now be applied to find an Euler tour of graph G*. An Euler tour of graph G* corresponds to a postman route in graph G that traverses each arc (i,j) $f(i,j) + 1$ times. Since the optimal values of $f(i,j)$ minimize expression (1), this Euler tour in graph G* must correspond to an optimal postman route in graph G.

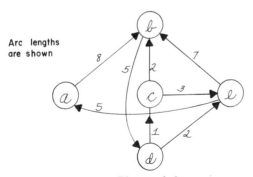

Figure 6.9

Postman Problem in a Directed Graph

EXAMPLE 1. Let us find an optimal postman route for the directed graph in Fig. 6.9. Examining the inner and outer degrees, we find that

$d^-(a) = 1 = d^+(a)$; vertex a is an intermediate vertex.
$d^-(b) = 3 > d^+(b) = 1$; vertex b is a source with $3 - 1 = 2$ units supply.
$d^-(c) = 1 < d^+(c) = 2$; vertex c is a sink with $2 - 1 = 1$ unit demand.
$d^-(d) = 1 < d^+(d) = 2$; vertex d is a sink with $2 - 1 = 1$ unit demand.
$d^-(e) = 2 = d^+(e)$; vertex e is an intermediate vertex.

Create a supersource S and join S to source vertex b by an arc (S,b) with capacity equal to the supply at source b, namely 2 units. Create a supersink T and join the sink vertices c and d to T by arcs (c,T) and (d,T). Let the capacity of arc (c,T) equal the demand at sink c, namely 1 unit. Let the capacity of arc (d,T) equal the demand at sink d, namely 1 unit. All other arc capacities are infinite. The resulting graph is shown in Fig. 6.10.

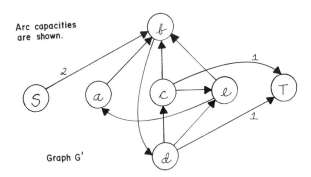

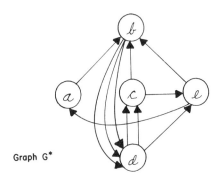

Figure 6.10

6.3 Postman Problem for Directed Graphs

At most two flow units can be sent from S to T, and all flow units must leave S by way of vertex b. At most one flow unit can arrive at T by way of vertex c, and at most one flow unit can arrive at T by way of vertex d. The paths taken by these flow units as they travel from S to T correspond to the arcs that the postman must repeat.

By performing the minimum cost flow algorithm or by close inspection, we can find that a minimum cost flow is 1 unit along (S,b), (b,d), (d,T), cost = 0 + 5 + 0 = 5; 1 unit along (S,b), (b,d), (d,c), (c,T), cost = 0 + 5 + 1 + 0 = + 6. Consequently, f(b,d) = 2, f(d,c) = 1, and all other f(i,j) = 0. Thus, the postman must repeat arc (b,d) twice and arc (d,c) once, adding an additional 5 + 5 + 1 = 11 units to his route.

Figure 6.10 shows graph G* which consists of all arcs in G together with two replicas of (b,d) and one replica of (d,c). Observe that graph G* is a symmetric graph, and as we know from Case A, graph G* possesses an Euler tour. For example, an Euler tour of graph G* and an optimal postman route of graph G is (a,b), (b,d), (d,c), (c,b), (b,d), (d,c), (c,e), (e,b), (b,d), (d,e), (e,a). The total length of this tour is 44 units, which is 33 units (the total length of all arcs in this graph) plus 11 units (which is the total length of all repeated arcs).

6.4 POSTMAN PROBLEM FOR MIXED GRAPHS

In this section, we shall consider the postman problem in a graph G in which some arcs are directed and some arcs are not directed (a mixed graph). If an arc is directed, then the postman must traverse this arc only along its direction (a one-way street). If an arc is not directed, then the postman may traverse this arc in either (or if necessary, both) directions (a two-way street). An undirected arc is not considered when calculating the inner and outer degrees of the vertices.

Can the postman always find a route in a mixed graph? Not always. It might happen that the graph contains a set S of vertices with the property that all arcs joining a vertex in S to a vertex not in S are directed towards the vertex in S. In this case, once the postman reaches a vertex in S he can never reach a vertex outside of S. He is trapped, and no solution exists for the postman problem. If no set S with this property exists, then the postman can keep on traveling, no matter how long it takes, until he has traversed all arcs and returned to his starting vertex.

For a mixed graph G = (X,A), three cases must be treated separately:

Case A: Graph G is even and symmetric.
Case B: Graph G is even but not symmetric.
Case C: Graph G is neither even nor symmetric.

Case A: This is the easiest case to analyze since the solution technique for this case is a composite of the solution techniques for even directed and even undirected graphs presented in Sec. 6.3 and 6.2, respectively.

Starting with any directed arc in graph G, generate a circuit of directed arcs as done in Sec. 6.3 for even directed graphs. (Since G is symmetric, this is possible.) Repeat this procedure until all directed arcs have been used. (This is possible since the unused arcs always form an even, symmetric graph.)

Next, repeat this procedure using only the undirected arcs in graph G. (Again, this is always possible, since the unused arcs form an even graph.) After all the arcs in graph G have been used, splice together all the circuits generated above into one circuit C. Circuit C forms an Euler tour of graph G and is an optimal solution to the postman problem for Case A.

Case B: Graph G is even but not symmetric. In this case, it is not easy to know in advance if the postman must repeat any arcs. For example, an Euler tour is an optimal solution for the even, nonsymmetric graph in Fig. 6.11, with undirected arc (a,b) traversed from a to b and undirected arc (c,a) traversed from c to a. On the other hand, no Euler tour can be optimal for the even, nonsymmetric

6.4 Postman Problem for Mixed Graphs

graph in Fig. 6.12 since arc (f,a) must be repeated twice so that the postman can exit vertex a along arcs (a,b) and (a,d), and also (a,e).

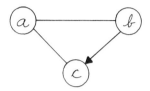

Figure 6.11

Mixed Graph with an Euler Tour

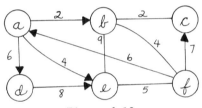

Figure 6.12

Mixed Graph with No Euler Tour

The mixed postman algorithm arbitrarily selects a direction for each undirected arc in graph G. This transforms graph G into an even directed graph G_D, and the solution technique for even, directed graphs given in Sec. 6.3 can be applied to graph G_D. However, due to the arbitrary choice of arc directions, some modifications are needed to correct some arc directions.

Mixed Postman Algorithm

Let G = (X,A) be any even, mixed graph. An optimal postman route (if one exists) can be found as follows.

Let U denote the set of all undirected arcs in graph G; let V denote the set of all directed arcs in graph G. Tentatively, select a direction for each arc in U. Call the resulting directed graph G_D. For each vertex i in G_D, calculate

$$\underline{D}(i) = d^-(i) - d^+(i) \tag{4}$$

If $\underline{D}(i) < 0$, then vertex i is a sink with demand equal to $-\underline{D}(i)$. If $\underline{D}(i) > 0$, then vertex i is a source with supply equal to $\underline{D}(i)$. If $\underline{D}(i) = 0$, then vertex i is an intermediate vertex.

If all vertices in G_D are intermediate, graph G_D is an even, symmetric directed graph, and the solution technique of Sec. 6.3 can be applied to graph G_D. This technique yields an Euler tour of G_D which corresponds to an optimal postman route of graph G.

Otherwise, construct a graph $G' = (X, A')$ as follows:

(a) For each arc $(i,j) \in V$, place an arc (i,j) in A' with infinite capacity and cost equal to the length of (i,j).

(b) For each arc $(i,j) \in U$, create two directed arcs (i,j) and (j,i) in A'. Let each of these arcs have infinite capacity and cost equal to the length of (i,j).

(c) For each arc $(i,j) \in U$, create a directed arc $(j,i)_1$ in A' whose direction is the reverse of the direction assigned this arc in G_D. These arcs are called *artificial* arcs. Assign each artificial arc a zero cost and a capacity equal to two.

Using the source supplies and sink demands defined above for graph G_D, apply the minimum cost flow algorithm to find a minimum cost flow in graph G' that satisfies all sink demands.

If no such flow exists, then no postman route exists. Otherwise, let $f(i,j)$ denote the number of flow units sent through arc (i,j) in G' in the minimum cost flow produced by the minimum cost flow algorithm. Recall from the minimum cost flow algorithm that each optimal flow value is a nonnegative integer. In the proof of this algorithm, it will be shown that each artificial arc carries either zero or two flow units.

Create a graph G^* as follows:

(a) For each nonartificial arc (i,j) in G' place $f(i,j) + 1$ copies of arc (i,j) in graph G^*

(b) If the flow in an artificial arc is two units, then place one copy of this arc in graph G^*

6.4 Postman Problem for Mixed Graphs

(c) If the flow in an artificial arc is zero, then reverse the direction of this arc and place one copy of this arc in graph G*. (Thus, if no units traverse an artificial arc, the tentative direction assigned to this arc in G_D is retained; if two flow units traverse an artificial arc, then the tentative direction assigned to this arc in G_D is reversed.)

Graph G* is an even, symmetric, directed graph. The solution technique for even, symmetric, directed graphs presented in Sec. 6.3 can now be applied to find an Euler tour of graph G*. This Euler tour of graph G* corresponds to an optimal postman route of the original graph G.

Proof: Since graph G is not necessarily symmetric, some vertices may have a surplus of incoming arcs; other vertices may have a surplus of outgoing arcs. Ideally, we would like to assign a direction to all the undirected arcs in G so that the resulting directed graph is symmetric. Then the solution technique for even, symmetric, directed graphs presented in Sec. 6.3 could be applied to find an Euler tour of graph G.

However, it can happen that there is no way to direct the undirected graphs so that the resulting graph is symmetric. In this case, some of the arcs (directed or undirected) must be repeated by the postman. Of course, the postman wants to select the repeated arcs so that their total length is as small as possible.

The algorithm selects a tentative direction for each undirected arc in G. The resulting directed graph is called G_D. The solution technique of Sec. 6.3 for even directed graph could be applied to graph G_D. However, the solution generated by this technique depends upon the tentative directions, and it is always possible that the tentative directions in graph G_D will lead to a nonoptimal solution to the postman problem.

The algorithm generates a graph G' and using the same source supplies and sink demands as in graph G_D finds a minimum cost flow that satisfies all these sink demands.

There are three kinds of arcs in graph G':

(a) Arcs corresponding to directed arcs in G (these arcs have nonzero cost and infinite capacity).

(b) Nonartificial arcs corresponding to undirected arcs in G (these arcs also have nonzero cost and infinite capacity).

(c) Artificial arcs corresponding to undirected arcs in G (these arcs have zero cost and a capacity equal to two).

The number $f(i,j)$ carried by an arc of type (a) or (b) in the minimum cost flow equals the number of times the corresponding directed arc is repeated by the postman. Thus, the postman will traverse each arc $(i,j) \in V$ $f(i,j) + 1$ times, and the postman will traverse each arc $(i,j) \in U$ a total of $f(i,j) + f(j,i) + 1$ times.

As shown later, each artificial arc [type (c)] carries either zero or two flow units. If an artificial arc $(j,i)_1$ carries two flow units in the minimum cost flow, this arc decreases the supply at j by two units and increases the supply at i by two units. This same effect could have been achieved by selecting the reverse tentative direction for this arc in graph G_D. Hence, the algorithm reverses the tentative direction of this arc.

If an artificial arc $(j,i)_1$ carries no flow units in the minimum cost flow, then this arc has no effect on the supplies at vertices i and j. This is equivalent to retaining the tentative direction given this arc in graph G_D. Hence, the algorithm retains the tentative direction given this arc.

From the minimum cost flow values $f(i,j)$ for the arcs in graph G', the algorithm generates a directed graph G*. It remains to show that

(a) Graph G* is even and symmetric

(b) An Euler tour of graph G* corresponds to an optimal postman route of graph G

(c) If the minimum cost flow algorithm can not find any flow that satisfied all sink demands in graph G', then no postman route exists for graph G.

6.4 Postman Problem for Mixed Graphs

Proof of (a): Since graph G_D is an even graph, $d^+(i) + d^-(i)$ is an even number, for all vertices i in G_D. Thus, $d^+(i)$ and $d^-(i)$ are both odd or are both even. In either case, $\underline{D}(i)$ must be even. Consequently, all supplies and demands in graph G' are even numbers (zero is an even number). Also all arc capacities are even numbers. Hence, the minimum cost flow algorithm will produce an optimal flow in which all flow values are even numbers. Thus, flow units in a minimum cost flow will travel in pairs.

The value of d(i) in G* equals the value of d(i) in G plus the number of flow units that enter or leave vertex i along nonartificial arcs. Since all flow units travel in pairs, it follows that d(i) is even in graph G*. Thus, graph G* is even.

A pair of flow units arriving at vertex i via a nonartificial arc increase $d^-(i)$ in G* by two units. A pair of flow units leaving vertex i along a nonartificial arc increase $d^+(i)$ by two units. A pair of flow units arriving at vertex i via an artificial arc cause the tentative direction of this arc to be reversed which has the effect of increasing $d^-(i)$ by two units. A pair of flow units leaving vertex i along an artificial arc cause the tentative direction of this arc to be reversed which has the effect of increasing $d^+(i)$ by two units. Thus, each flow unit arriving at vertex i increases $d^-(i)$ by one unit, and each flow unit leaving vertex i increases $d^+(i)$ by one unit. Since the minimum cost flow satisfies equation (3) for all vertices in graph G, it follows that graph G* is symmetric.

Proof of (b): Suppose that the postman route produced by the mixed postman algorithm is not optimal. Then, there exists another postman route whose duplicated arcs have an even smaller total length. This route must correspond to a flow in graph G' with even lower cost than the flow generated by the minimum cost flow algorithm, which is a contradiction.

Proof of (c): Since $\sum_i d^-(i) = \sum_i d^+(i)$ in any graph G_D, the total of the source supplies in graph G_D must equal the total of the sink demands in graph G_D. Thus, all source supplies must be shipped out in order to satisfy all sink demands.

If graph G' contains an arc from i to j, then graph G' must contain an arc with infinite capacity from i to j. Suppose that the minimum cost flow algorithm terminates without satisfying all sink demands. Let S denote the set of all vertices that were colored after the last iteration of the minimum cost flow algorithm. From the flow augmenting algorithm, which is a subroutine of the minimum cost flow algorithm, we know that all arcs from a vertex in S to a vertex not in S must carry a full capacity flow. This is impossible since some of these arc capacities are infinite. Thus, no such arcs can exist, and all arcs with one colored endpoint and one uncolored endpoint are directed into set S. Consequently, once the postman reaches set S he cannot leave set S and no postman route is possible. Q.E.D.

EXAMPLE. Let us find an optimal postman route for the even, non-symmetric, mixed graph G shown in Fig. 6.12. First, directions are arbitrarily selected for each undirected arc in graph G. The resulting graph G_D is shown in Fig. 6.13. In graph G_D

$d^+(a) = 3$, $d^-(a) = 1$, $\underline{D}(a) = -2$; vertex a is a sink with demand 2
$d^+(b) = 2$, $d^-(b) = 2$, $\underline{D}(b) = 0$; vertex b is an intermediate vertex
$d^+(c) = 0$, $d^-(c) = 2$, $\underline{D}(c) = 2$; vertex c is a source with supply 2
$d^+(d) = 1$, $d^-(d) = 1$, $\underline{D}(d) = 0$; vertex d is an intermediate vertex
$d^+(e) = 2$, $d^-(e) = 2$, $\underline{D}(e) = 0$; vertex e is an intermediate vertex
$d^+(f) = 2$, $d^-(f) = 2$, $\underline{D}(f) = 0$; vertex f is an intermediate vertex.

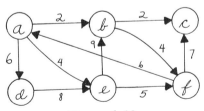

Figure 6.13
Graph G_D

6.4 Postman Problem for Mixed Graphs

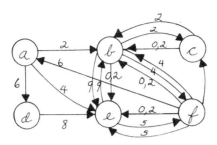

Figure 6.14
Graph G'

Graph G' is shown in Fig. 6.14. The first number next to each arc denotes its cost. If the arc capacity is finite, it is denoted by the second number next to the arc.

The minimum cost flow algorithm is now required to find a minimum cost way of sending 2 flow units from source c to sink a. (Notice that all supplies and demands are even numbers and that the total supply at the sources equals the total demand at the sinks.) By inspection, we can see that the minimum cost flow consists of sending both flow units from c to a along the path $(c,b)_1$, (b,f), (f,a). The total cost is $0 + 4 + 6 = 10$ units for each flow unit, or 20 units. Thus, the optimal flow values are $f(c,b)_1 = f(b,f) = f(f,a) = 2$, and all other flow values equal zero. Since $f(c,b)_1 = 2$, we must reverse the tentative direction of this arc so that it is directed from c to b. All other arbitrary directions are retained. Moreover, arcs (b,f) and (f,a) must be repeated twice since $f(b,f) = f(f,a) = 2$.

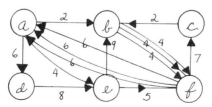

Figure 6.15
Graph G*

Graph G* is shown in Fig. 6.15. Note that graph G* is an even, symmetric directed graph. The solution technique for even, symmetric directed graphs given in Sec. 6.3 can now be applied to find an Euler tour of graph G*. This tour corresponds to an optimal postman route in graph G. Since the minimum cost flow in G' cost 20 units, the optimal postman route in graph G will repeat arcs with a total length of 20 units.

Case C: Graph G is neither even nor symmetric.

As far as the author knows, no optimal solution technique is currently available for this case.

This problem could be approached as a two-stage problem: First make the graph even in an optimal way; second make the even graph symmetric using the optimal procedure given in Sec. 6.4, Case B. However, there is no guarantee that optimally solving each stage will lead to an optimum solution for the overall problem.

EXERCISES

1. The city administration has declared that from now on a certain two-way street shall be a one-way street. Will this necessarily increase the length of the optimal route of the postman who must serve this street? Describe conditions under which this change will increase the length of the postman's optimal route.
2. The city administration has declared that from now on a certain one-way street shall be a two-way street. Will this necessarily decrease the length of the optimal route of the postman who must serve this street? Describe conditions under which this change will decrease the length of the postman's optimal route.
3. Find an optimal postman route for the graph shown in Fig. 6.16.
4. Prove that the postman need never repeat an arc in an undirect graph more than once. Is this also true for directed graphs?
5. When solving the postman problem for an undirected graph, show that at most k iterations of the Floyd or Dantzig algorithm are ever needed, where k is the number of even degree vertices in the graph. (Hint: an odd-degree vertex will never be an intermediate vertex in any path of repeated edges.)

Exercises 259

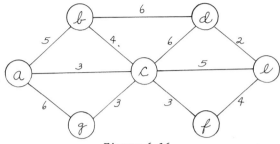

Figure 6.16

6. Find an optimal postman route for the directed graph shown in Fig. 6.13.
7. How much walking can the postman save by changing starting points?
8. Suppose that the postman wishes to start his route at one vertex and finish at a different vertex. How can this be incorporated into the postman problem?
9. Suppose that the postman must traverse street A before traversing street B. How can this be incorporated into the postman problem for
    (a) An undirected graph
    (b) A directed graph
    (c) A mixed graph
10. Solve the postman problem for the mixed graph shown in Fig. 6.12 using tentative directions different from those selected for the example that used this graph.
11. In the minimum cost flow produced by the minimum cost flow algorithm to solve the mixed postman problem, all flow units travel in pairs. Suppose that another minimum cost flow exists and in this flow not all flow units travel in pairs. How should you treat an artificial arc with only one flow unit on it? (Answer: Let this arc be undirected in graph G*.)
12. Can the Königsberg bridge problem described in Sec. 1.1 be solved without repeating any bridges?

13. A newsboy must deliver papers on both sides of both Campbell Avenue and Maplewood Avenue between 63rd Street and 66th Street. The number of papers to be delivered on each side of each street is in the following diagram. It takes the newsboy 1 minute to pedal one block plus 6 seconds for each paper he tosses on a porch. He must start and finish his route at the corner of 63rd and Maplewood. What is the best route for the newsboy?

	0	8	4	
		Campbell Avenue		
	3	0	6	
	6	0	3	
		Maplewood Avenue		(1/2 Block)
Station	x 5	0	0	
	63rd St.	64th St.	65th St.	66th St.

|←(1 Block)→|

REFERENCES

Edmonds, J., and Ellis L. Johnson, 1973. Matching, Euler Tours and the Chinese Postman, *Math. Prog.*, vol. 5, pp. 88-124. (This paper provides an excellent treatment of postman results, additional methods for generating Euler tours and a long bibliography for this problem.)

Chapter 7

TRAVELING SALESMAN PROBLEM

7.1 SALESMAN PROBLEMS

A traveling salesman is required to call at each town in his district before returning home. Needless to say, the salesman would like to route his calls so as to travel as little as possible. Thus, the salesman encounters the problem of finding a route that minimizes the total distance (or time or cost) needed to visit all the towns in his district.

The traveling salesman problem can be rephrased in terms of a graph: Construct a graph $G = (X,A)$ whose vertices correspond to the towns in the salesman's district and whose arcs correspond to the roads joining two towns. Let the length $a(x,y) \geq 0$ of each arc $(x,y) \in A$ equal the length (or time or cost) of the corresponding journey along arc (x,y). A circuit that includes each vertex in graph G at least once is called a *salesman circuit*. A circuit that includes each vertex in graph G exactly once is called a *Hamiltonian circuit* after the Irish mathematician Sir William Rowan Hamiltonian who first studied these problems in 1859. The *general salesman problem* is the problem of finding a salesman circuit with the smallest possible total length. The *salesman problem* is the problem of finding a Hamiltonian circuit with the smallest possible total length.

EXAMPLE 1. A bank courier must deliver letters from his home branch to every other branch bank every day. Usually, when he is delivering at a branch bank, he is imposed upon to carry some additional letters to his next stop. Being adverse to additional work, the courier would like to know how he should arrange his stops so as to minimize the total number of additional letters he must carry.

The bank courier solves his problem by solving the general salesman problem on the graph whose vertices correspond to the branch banks and whose arcs correspond to the possible trips between branch banks. Let the length of an arc (x,y) in this graph equal the predicted number of additional letters that the courier would be asked to carry from branch x to branch y.

Suppose the bank courier is desirous of promotion and wishes to maximize the total number of additional letters carried. The courier could, of course, circulate between the banks forever and ultimately carry an infinite number of letters. However, suppose that he is permitted to visit each bank only once. Let M equal some very large number, and let each arc length now equal M less its original length. If there are n banks in the system, each Hamiltonian circuit of the graph consists of n arcs. Now, the courier's problem is solved by finding a smallest total length Hamiltonian circuit.

EXAMPLE 2. In a machine shop, a job must be run through each of n different machines in no particular sequence. However, a set-up time of a(x,y) is required whenever a job goes from machine x to machine y. What is the fastest way to route a job through each of the n machines?

This problem is solved by solving the salesman problem on the following graph G. Let each vertex correspond to a machine. Let each pair (x,y) of vertices in graph G be joined by an arc with length a(x,y).

A salesman circuit with least total length is called an *optimum salesman circuit* and is an optimum solution for the general salesman problem. A Hamiltonian circuit with least total length is called an *optimum Hamiltonian circuit* and is an optimum solution to the salesman problem.

An optimum salesman circuit need not be an optimum Hamiltonian circuit. For example, consider the graph shown in Fig. 7.1. The

7.1 Salesman Problems

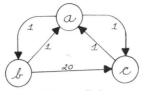

Figure 7.1

Optimum Salesman Route

only Hamiltonian circuit in this graph is (a,b), (b,c), (c,a) which has a total length equal to 1 + 20 + 1 = 22 units. The (optimum) salesman circuit (a,b), (b,a), (a,c), (c,a) that passes through vertex a twice has total length equal to 1 + 1 + 1 + 1 = 4 units. Thus, an optimum salesman circuit need not be an optimum Hamiltonian circuit.

When is the solution to the general salesman problem a Hamiltonian circuit?

THEOREM 7.1 If for each pair x,y of vertices in graph G,

$$a(x,y) \leq a(x,z) + a(z,y) \quad \text{(for all } z \neq x, z \neq y\text{)} \quad (1)$$

then a Hamiltonian circuit is an optimum solution (if a solution exists) to the general salesman problem for graph G.

Condition (1) merely says that the direct distance from x to y is never more than the distance via any other vertex z. Condition (1) is called the *triangle inequality*.

Proof: Suppose that no optimum solution to the general salesman problem is a Hamiltonian circuit. Let C be any optimum salesman circuit. Since C is not a Hamiltonian circuit, then some vertex, say vertex z, appears at least twice in circuit C. Suppose that the first time the salesman arrives at vertex z he arrives from vertex x and departs to vertex y. Alter circuit C so that the salesman travels from x directly to y bypassing z. The resulting route C' is also a circuit since it visits every vertex at least once. Moreover, by (1), the total length of C' does not exceed the length of C. Replacing C by C' and repeating this argument, we generate another salesman circuit C", etc. Eventually, this process leads us to an optimum circuit that is Hamiltonian since each successive circuit has one less arc than its predecessor. Q.E.D.

From Theorem 7.1, it follows that if graph G satisfies the triangle inequality, then the optimum solutions for the salesman problem for graph G are optimum solutions for the general salesman problem in graph G.

There is a simple way to spare outselves the needless trouble of developing two solution techniques, one for the general salesman problem and one for the salesman problem. If graph G does not satisfy the triangle inequality then replace each arc length a(x,y) that fails the triangle inequality with the length of a shortest path from x to y. Record that the arc from x to y no longer represents a direct journey from x to y but now represents a journey along a shortest path from x to y. Now, a(x,y) just satisfy the triangle inequality.

If an optimum solution to the salesman problem for graph G contains an arc (x,y) whose length was shortened as specified above, then replace arc (x,y) by a shortest path from x to y in the optimum solution. *Thus, we need solution techniques for only the salesman problem.*

For example, in Fig. 7.1, a(b,c) = 20 > a(b,a) + a(a,c) = 1 + 1. Thus, (1) fails for arc (b,c). If the length of arc (b,c) is reduced to 2, the length of a shortest path from b to c, then the only Hamiltonian circuit in the resulting graph is (a,b), (b,c), (c,a) whose length is 1 + 2 + 1 = 4. Replacing (b,c) by (b,a), (a,c) yields the circuit (a,b), (b,a), (a,c), (c,a) which is an optimum salesman circuit for the original graph.

Suppose that instead of wishing to find a Hamiltonian circuit with the smallest total length, we wanted to find a Hamiltonian circuit with the *largest* total length. For example, the bank courier might want to maximize rather than minimize the total number of additional letters that he carries. Can this problem also be solved as a salesman problem?

Let M denote the largest arc length in graph G. Let

$$a'(x,y) = M - a(x,y) \geq 0 \quad [\text{for all } (x,y)] \qquad (2)$$

Every Hamiltonian circuit in graph G = (X,A), contains exactly $|X|$ arcs. An optimum (shortest) Hamiltonian circuit C based on the primed

7.1 Salesman Problems

arc lengths has total length equal to

$$\sum_{(x,y) \in C} a'(x,y) = \sum_{(x,y) \in C} [M - a(x,y)]$$
$$= |X|M - \sum_{(x,y) \in C} a(x,y)$$

Thus, it follows that a minimum length Hamiltonian circuit based on the primed arc lengths corresponds to a maximum length Hamiltonian circuit based on the original unprimed arc lengths.

Thus, to find a maximum length Hamiltonian circuit, we need only solve the salesman problem using arc lengths transformed as in (2).

Not all graphs possess a Hamiltonian circuit. For example, the graph consisting solely of two vertices x and y and a single arc (x,y) does not contain a Hamiltonian circuit. Section 7.2 describes conditions that insure that a graph possesses a Hamiltonian circuit. Section 7.3 presents methods for calculating a lower bound on the length of an optimum Hamiltonian circuit. Section 7.4 describes techniques for finding an optimum Hamiltonian circuit.

7.2 EXISTENCE OF A HAMILTONIAN CIRCUIT

As shown in Sec. 7.1, the salesman problem is solved by finding an optimum Hamiltonian circuit. Unfortunately, not all graphs contain a Hamiltonian circuit. Consequently, before proceeding to look for an optimum Hamiltonian circuit, we should at least try to establish if the graph possesses any Hamiltonian circuits. This section describes several conditions under which a graph possesses a Hamiltonian circuit. A very extensive treatment of existence conditions for Hamiltonian circuits can be found in Berge (1973).

A graph is called *strongly connected* if for any two vertices x and y in the graph, there is a path from x to y. A subset X_i of vertices is called a *strongly connected vertex subset* if for any two vertices $x \in X_i$ and $y \in X_i$, there is a path from x to y in the graph and X_i is contained in no other set with the same property. The subgraph generated by a strongly connected vertex subset is called a *strongly connected component* of the original graph.

For example, the graph in Fig. 7.1 is strongly connected since there is a path from every vertex to every other vertex. The reader should verify this. Next, consider the graph in Fig. 7.2. This graph is not strongly connected because there is no path from vertex d to vertex b, although there is a chain from d to b, namely arc (d,b). The vertices {a,b,c} form a strongly connected vertex subset since there is a path from each of these vertices to every other vertex in this set. Moreover, no other vertex can be added to this set without losing this property. For example, vertex d cannot be added to the set since there is no path from d to a. The subgraph generated by {a,b,c} is shown in Fig. 7.2. This subgraph is a strongly connected component of the original graph.

There is a path from d to e and a path from e to d. However, {d,e} is not a strongly connected vertex subset because vertex f can be added to this set without losing the strongly connected property. No other vertices can be added without losing this property. Hence {d,e,f} is a strongly connected vertex subset. The strongly connected component generated by {d,e,f} is also shown in Fig. 7.2.

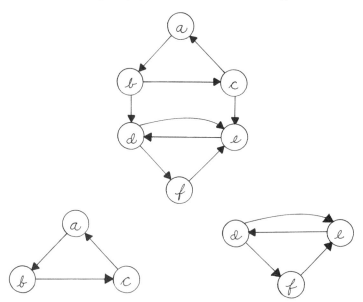

Figure 7.2
Graph and its Strongly Connected Components

7.2 Existence of a Hamiltonian Circuit

If graph G is not strongly connected, then graph G does not possess a Hamiltonian circuit. This follows since a Hamiltonian circuit contains a path between each pair of vertices in the graph. Thus, a necessary condition for the existence of a Hamiltonian circuit is that graph G be strongly connected.

Recall that a loop is any arc whose head and tail are the same vertex, i.e., an arc of the form (x,x). No Hamiltonian circuit can contain a loop, and consequently, the existence of a Hamiltonian circuit is not affected by adding or deleting loops from graph G.

If two vertices, say x and y, are joined by more than one arc $(x,y)_1$, $(x,y)_2$, ..., with the same direction, then the deletion of all of these arcs, except the shortest one, does not effect the existence of a Hamiltonian circuit in the graph or the length of an optimum Hamiltonian circuit, (if one exists).

For these reasons, we shall henceforth assume that graph G contains no loops and not more than one arc from x to y for all x and y.

Let $D^-(x)$ denote the set of all vertices y in graph G = (X,A) such that $(y,x) \in A$, i.e., the set of all vertices "incident into" vertex x. Let $D^+(x)$ denote the set of all vertices y such that $(x,y) \in A$, i.e., the set of all vertices "incident from" vertex x. The vertices in $D^-(x)$ and $D^+(x)$ are respectively called the *predecessors* and *successors* of vertex x. Let $D(x) = D^-(x) \cup D^+(x)$. As before, let $d^-(x) = |D^-(x)|$ and $d^+(x) = |D^+(x)|$ and $d(x) = |D(x)|$. Lastly, let n denote the number of vertices in graph G.

With these definitions in mind, we can now state the following very general theorem due to Ghouila-Houri (1960):

THEOREM 7.2 If graph G = (X,A) satisfies the following conditions, then graph G possesses a Hamiltonian circuit:

(I) Graph G is strongly connected
(II) $d(x) \geq n$, for all $x \in X$.

Proof: The proof is achieved by induction on n, the number of vertices in graph G. Trivially, the theorem is true for n = 2 and n = 3. We shall assume that the theorem is true for all graphs with less than n (n > 3) vertices and use this assumption to show that the theorem is also valid for all graphs with n vertices.

Let G be any graph with n vertices that satisfies conditions (I) and (II). Let C denote a simple circuit in G with the largest possible number of arcs. Let x_1, x_2, ..., x_m denote the sequence in which C visits the vertices of G. If m = n, the theorem is true. Otherwise, m < n and C is not a Hamiltonian circuit. Let $X_0 = \{x_1, x_2, ..., x_m\}$. Let X_1, X_2, ..., X_p denote the strongly connected components of the subgraph generated by $X - X_0$.

Claim (a): Each strongly connected component X_1, X_2, ..., X_p contains a Hamiltonian circuit.

To prove Claim (a), we need only show that the degree in X_i of each vertex $x \in X_i$ is at least $|X_i|$ for all i = 1, 2, ..., p. Then component X_i will satisfy both conditions (I) and (II), and by the induction hypothesis contain a Hamiltonian circuit.

We know that $d(x) \geq n$. Consider any vertex $y \in X_j$, $j \neq i$. Both arcs (x,y) and (y,x) cannot be present in A; otherwise, X_i and X_j would form one strongly connected component. For k = 1, 2; ..., m, both (x_k,x) and (x,x_k) cannot exist, otherwise circuit C could be extended to include vertex x. Consequently, the number of arcs joining x to a vertex not in X_i cannot exceed $|X| - |X_i|$. Thus, the degree of x in X_i is not less than X_i, and Claim (a) is true.

Claim (b): For i = 1, 2, ..., p,

$$|X_i| \leq |X_0|$$

Otherwise, there would exist a Hamiltonian circuit in X_i that contains more arcs than circuit C.

Claim (c): At least $2 + |X_0| - |X_i| \geq 2$ arcs join each vertex $x \in X_i$ to vertices in X_0.

As mentioned in Claim (a), not more than $|X_j|$ arcs join vertex x and vertices in X_j for all $j \neq i$, $j \neq 0$. Also, not more than $2(|X_i| - 1)$ arcs can join vertex x to vertices in X_i. Since $d(x) \geq n$, it follows that at least $2 + |X_0| - |X_i|$ arcs join vertex x to vertices in X_0 and (c) follows.

7.2 Existence of a Hamiltonian Circuit

Claim (d): There exists a component X_i such that there is an arc from a vertex in X_i to a vertex in X_0 and an arc from a vertex in X_0 to a vertex in X_i.

For $p = 1$, Claim (d) must be true since G is strongly connected.

For $p > 1$, suppose that all arcs joining X_1 and X_0 are directed from X_1 to X_0. Consider the subgraph generated by $X_1 \cup X_0$. Since a vertex in X_1 is joined at most once to each vertex not in $X_1 \cup X_0$; the vertices in X_1 satisfy condition (II) in this subgraph. It follows from Claim (a) that the vertices in X_0 also satisfy condition (II) in this subgraph.

Since G is strongly connected, there is a path P from vertex $x \in X_0$ to a vertex $y \in X_1$ such that the intermediate vertices of this path are not in $X_1 \cup X_0$. Path P must contain at least two arcs.

If arc (x,y) were added to the subgraph generated by $X_1 \cup X_0$, then this subgraph would be strongly connected. Consequently, this subgraph together with arc (x,y) would satisfy the induction hypothesis, and there would exist a Hamiltonian circuit of this subgraph that contains arc (x,y). Replace arc (x,y) with path P. The resulting circuit contains more arcs than circuit C, which is impossible. Hence, each component X_1, X_2, ..., X_p must be joined to X_0 by arcs in both directions. This proves Claim (d).

Claim (e): Let X_1 be a component satisfying Claim (d). For each vertex $y \in X_1$, there is an arc from y to a vertex in X_0 and there is an arc from a vertex in X_0 to vertex y.

Let C_1 be a Hamiltonian circuit of X_1. Let y_1, y_2, ..., y_q denote the order in which C_1 visits the vertices of X_1. (Circuit C_1 exists from Claim (a).) Suppose that there is no arc from any vertex in X_0 to vertex y_j. Follow circuit C_1 through the vertices y_{j+1}, y_{j+2}, ..., until a vertex y_{j+t} is encountered such that there is an arc (x_k, y_{j+t}) originating in X_0. Consider vertex y_{j+t-1}. No arc can be directed from y_{j+t-1} to x_{k+1}, x_{k+2}, ..., x_{k+q}. Otherwise, circuit C would be merged with the q - 1 arcs from y_{j+t} to y_{j+t-1} to create a circuit with more than m arcs, which is impossible. Consequently, no arcs are directed from X_0 into y_{j+t-1} and at most

$m - q$ arcs are directed from y_{j+t-1} to vertices in X_0. Since $m = |X_0|$ and $q = |X_1|$, this contradicts Claim (c), and Claim (e) must be valid.

Claim (f) (Conclusion): As before, let $x_1, x_2, \ldots, x_m$ denote the order of the vertices visited by C, and let $y_1, y_2, \ldots, y_q$ denote the order of the vertices visited by C_1. Let

$$a(i,j) = \begin{cases} 1 & \text{If } (x_i, y_j) \in A \\ 0 & \text{Otherwise} \end{cases}$$

$$b(i,j) = \begin{cases} 1 & \text{If } (y_{j-1}, x_{i+1}) \in A \\ 0 & \text{Otherwise} \end{cases}$$

Since there are at least $q(m - q + 2)$ arcs joining X_0 and X_1 from Claim (c), it follows that

$$\sum_{i,j} a(i,j) + b(i,j) \geq q(m - q + 2)$$

For some j_0, it follows that

$$\sum_i a(i,j_0) + b(i,j_0) \geq m - q + 2 \tag{3}$$

From Claim (e), there is at least one arc (x_{i_0}, y_{j_0}) directed from X_0 into vertex y_{j_0}. Consequently, since no circuits with more than m arcs exist, it follows that $b(i_0, j_0) = 0$, $b(i_0 + 1, j_0) = 0$, $\ldots$, $b(i_0 + q - 1, j_0) = 0$. Thus, there are at most $m - q + 1$ arcs from y_{j_0-1} to vertices in X_0. If $a(i_0 + 1, j_0) = 1$, then $b(i_0 + q, j_0) = 0$. If $a(i_0 + 2, j_0) = 1$, then $b(i_0 + q + 1, j_0) = 0$, etc. Thus, the existence of each additional arc directed from X_0 to y_{j_0} prohibits the existence of at least one more arc from y_{j_0} to X_0.

Consequently,

$$\sum_i a(i,j_0) + b(i,j_0) \leq m - q + 1$$

which contradicts inequality (3). Thus, circuit C cannot contain only $m < n$ arcs, which is a contradiction. Q.E.D.

7.2 Existence of a Hamiltonian Circuit

It is easy to verify that a graph satisfies conditions (I) and (II) required by Theorem 7.2. Condition (I) is verified by applying either the Floyd or Dantzig shortest path algorithm (see Chap. 3) to ascertain if there is a path with finite length joining every pair of vertices in the graph. Condition (II) is verified simply by counting the number of arcs incident to each vertex in the graph.

If graph G is an undirected graph, then no direction is specified on each arc and the salesman is allowed to traverse an arc in either direction. In this case, the salesman problem is solved by finding an *optimum Hamiltonian cycle*, i.e., a shortest simply cycle that contains all the vertices of the graph. The following result due to Chvàtal (1972) describes a sufficient condition for an undirected graph to possess a Hamiltonian cycle.

Again, let n denote the number of vertices in the graph. Name the vertices $x_1, x_2, \ldots, x_n$ so that $d(x_1) \leq d(x_2) \leq \ldots \leq d(x_n)$

THEOREM 7.3 If $n \geq 3$, and if

$$d(x_k) \leq k < 1/2 \, n \Rightarrow d(x_{n-k}) \geq n - k \qquad (4)$$

then graph $G = (X, E)$ contains a Hamiltonian cycle.

Proof: Supeose that the theorem is false, i.e., graph G satisfies condition (4) but contains no Hamiltonian cycle. Consider any edge (x_i, x_j) not present in G. If the addition of this edge does not create a Hamiltonian cycle, add this edge to graph G. Repeat this process until no more edges can be added to graph G. Note that condition (4) remains satisfied after the addition of each new edge to graph G since the additional edge does not lower the degree of any vertex. Call the final graph $G^* = (X, E^*)$. We shall now use graph G^* to obtain a contradiction.

Let u and v be any two nonadjacent vertices such that $d(u) + d(v)$ is as large as possible. Without loss of generality, assume that $d(u) \leq d(v)$. Let C denote the longest simple chain from u to v. Since no additional edges can be added to G^* without creating a Hamiltonian cycle, it follows that C contains all the vertices in X. Let $u = u_1, u_2, \ldots, u_{n-1}, u_n = v$ denote the order in which C visits the vertices of X.

Let S denote the set of all vertices u_i such that u_{i+1} is adjacent to u. Let T denote the set of all vertices adjacent to v. No vertex u_j can be a member of both S and T since otherwise, the cycle u_j, u_{j-1}, ..., u_1, u_{j+1}, u_{j+2}, ..., u_n, u_j would be a Hamiltonian cycle. Also, S ∪ T = $\{u_1, u_2, ..., u_{n-1}\}$. Therefore, d(u) + d(v) = $|S| + |T| \leq n$. Consequently, d(u) $\leq$ 1/2 n.

Since no vertex u_j is a member of both S and T, if $u_j \in$ S, then u_j is not adjacent to v. By the maximality of d(u) + d(v) it follows that $d(u_j) \leq d(u)$. Thus, there are at least $|S|$ vertices whose degree do not exceed the degree of u. Set k = d(u). Thus, $d(x_k) \leq$ k. By condition (4), it follows that $d(x_{n-k}) \geq$ n - k. Thus, there must be at least k + 1 vertices whose degree is at least n - k. Since d(u) = k, vertex u is not adjacent to all of these k + 1 vertices. Thus, there is a vertex w that is not adjacent to u and d(w) $\geq$ n - k. Consequently, d(u) + d(w) > d(u) + d(v), which is a contradiction. Consequently, graph G* must contain a Hamiltonian cycle. Q.E.D.

Condition (4) is easy to verify. Merely, order the vertices according to ascending degree and check if condition (4) is satisfied for the first 1/2 n vertices.

7.3 LOWER BOUNDS

This section presents methods for calculating lower bounds on the length of an optimum Hamiltonian circuit in graph G = (X,A). If a lower bound is subtracted from the length of any Hamiltonian circuit, the difference equals the maximum amount by which this Hamiltonian circuit can exceed optimality. This is useful when evaluating non-optimum Hamiltonian circuits.

The arcs of a Hamiltonian circuit must satisfy two properties:

1. Each vertex must have one arc incident to it and one arc incident out of it
2. The arcs are connected.

Any set of disjoint circuits that contain all vertices satisfy item 1. Any connected set of arcs satisfy item 2. Only Hamiltonian circuits satisfy both items 1 and 2.

7.3 Lower Bounds

Consider the family $\underline{F}$ of all subsets of arc set A that satisfy property 1. (We shall assume that $\underline{F}$ is not empty.) Clearly, every Hamiltonian circuit is a member of $\underline{F}$. Thus, the length of the shortest member of $\underline{F}$ is a lower bound on the length of an optimum Hamiltonian circuit. Denote this lower bound by L_1.

Lower bound L_1 can be calculated as follows:

Step 1: Construct a graph $G' = (X', A')$ as follows: For each vertex $x \in X$, create two vertices $x_1 \in X'$ and $x_2 \in X'$. For each arc $(x,y) \in A$, create a "middle" arc $(x_1, y_2) \in A'$. Let each middle arc have capacity equal to one and cost equal to the length of the corresponding arc in A. Create a source vertex s and a sink vertex t. Create a source arc (s, x_1) from s to each vertex $x_1 \in A'$. Create a sink arc (x_2, t) from each vertex $x_2 \in A'$ to t. Let each source arc and sink arc have capacity equal to one and cost equal to zero. (see Fig. 7.3).

Step 2: Send as many flow units from s to t in G' as possible with minimum total cost. This is accomplished by using the minimum cost flow algorithm (see Sect. 4.3). Let L_1 equal the cost of the resulting flow.

Since all arc capacities in G' equal one, the resulting flow consists of flow units traveling alone from s to t. Moreover, no intermediate vertex (i.e., no vertex except s and t) comes in contact with more than one flow unit.

Each middle arc in A' corresponds to an arc in A. Consider the set of all middle arcs that carry one flow unit. Let M denote the corresponding set of arcs in A. The arcs in set M form disjoint circuits in G that contain all the vertices of G. If this were not the case, then some vertex $x \in X$ would have no arc of M incident into it or no one arc of M incident out of it. The former implies that no flow unit enters vertex x_2; the latter implies that no flow unit leaves vertex x_1. In either case, the total flow from s to t would consist of less than $|X|$ flow units. But this is impossible since each member of $\underline{F}$ corresponds to a flow in G' of $|X|$ units and set $\underline{F}$ is not empty.

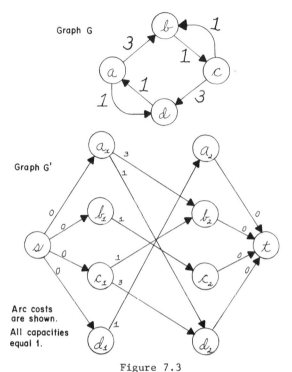

Figure 7.3
Calculating Lower Bounds

Since only middle arcs have nonzero costs, the total cost of the resulting flow equals the total length of the corresponding member of $\underline{F}$. Since the resulting flow has minimum total cost, the corresponding member of $\underline{F}$ must be the minimum member of $\underline{F}$. Thus, L_1 is a lower bound on the length of the optimum Hamiltonian circuit in G.

If graph G contains undirected arcs, each undirected arc can be replaced by two oppositely directed arcs joining the same vertices. Let each of these directed arcs have length equal to the length of the original undirected arc. The resulting graph will contain only directed arcs, and the technique described above for finding L_1 can be applied to the resulting graph.

If set M corresponds to a Hamiltonian circuit, then we have not only generated a lower bound on the length of an optimum

7.3 Lower Bounds

Hamiltonian circuit, but we have also generated an optimum Hamiltonian circuit.

The reader familiar with linear programming techniques will note that bound L_1 could have also been found by using the assignment algorithm. A technique for improving bound L_1 has been developed by Christofides (1972). This technique uses the assignment algorithm.

EXAMPLE 1. Compute a lower bound on the length of an optimum Hamiltonian circuit for the graph G shown in Fig. 7.3. Graph G' is also shown in Fig. 7.3. Obviously, not more than four units can flow from s to t in graph G' since the capacity of all the arcs directed into sink t equals 4. Each middle arc costs at least one unit. Since graph G' is small, we can find the minimum cost flow by inspection rather than by using the minimum cost flow algorithm. The minimum cost flow is

Quantity	Path	Cost
1 unit	(s,a_1), (a_1,d_2), (d_2,t)	$0 + 1 + 0 = 1$
1 unit	(s,b_1), (b_1,c_2), (c_2,t)	$0 + 1 + 0 = 1$
1 unit	(s,c_1), (c_1,b_2), (b_2,t)	$0 + 1 + 0 = 1$
1 unit	(s,d_1), (d_1,a_2), (a_2,t)	$0 + 1 + 0 = 1$

The total cost of this flow is 4; consequently, $L_1 = 4$. The arcs in G that correspond to the middle arcs in G' that carry flow are (a,d), (d,a), (b,c), (c,b). These arcs form two circuits (a,d), (d,a), and (b,c), (c,b). Observe that these circuits satisfy property 1 but not property 2. This set of arcs is a member of $\underline{F}$. The unique Hamiltonian circuit in graph G is (a,b), (b,c), (c,d), (d,a) with total length of $3 + 1 + 3 + 1 = 8$.

Now, let us compute another lower bound L_2 on the length of an optimum Hamiltonian circuit in graph G. This lower bound is calculated by relaxing property 1 and not specifying how many arcs leave each vertex.

Recall from Chap. 2 that an *arborescence* is a tree in which not more than one arc is directed into any vertex. The unique vertex in an arborescence that has no arc directed into it is called the *root* of the arborescence. Consider any set of arcs that is a spanning arborescence of graph G rooted at some vertex, say vertex x, together with an arc directed into vertex x. Let $\underline{H}$ denote the family of all such sets of arcs (Assume set $\underline{H}$ is not empty.) Each member of $\underline{H}$ satisfies property 2 but not necessarily property 1. Every Hamiltonian circuit of G is a member of $\underline{H}$.

The maximum branching algorithm of Chap. 2 can be applied to find a minimum total length spanning arborescence of graph G rooted at vertex x. Add to this arborescence the shortest arc directed into root x. Call the resulting set of arcs T. Set T must be a minimum length member of $\underline{H}$; consequently, the total length of the arcs in set T generates a lower bound on the length of an optimum Hamiltonian circuit in graph G. Call this lower bound L_2.

If set T forms a Hamiltonian circuit, then this circuit is an optimum Hamiltonian circuit.

EXAMPLE 2. Let us generate the lower bound L_2 for the length of an optimum Hamiltonian circuit for the graph in Fig. 7.3. Since the graph is small, we can find by inspection rather than by resorting to the maximum branching algorithm a minimum spanning arborescence rooted at vertex a. It is (a,b), (b,c) and (a,d) with total length $3 + 1 + 1 = 5$. The shortest arc incident into vertex a is arc (d,a) with length 1. Thus, set T consists of the four arcs (a,b), (b,c), (a,d), (d,a), whose total length is $3 + 1 + 1 + 1 = 6$. Thus, $L_2 = 6$. Note that T does not form a Hamiltonian circuit.

Recall from the previous example that $L_1 = 4 \neq L_2$. Thus, L_1 and L_2 are not always equal.

If graph G is undirected, then the procedure for calculating L_2 can be simplified: simply remove any vertex x from the graph. Use the maximum spanning tree algorithm of Chap. 2 to find a minimum spanning tree of the remaining graph. Select the two shortest arcs incident to vertex x, and add these two arcs to the spanning tree. Call the resulting set of edges U.

Set U satisfies property 2 but not necessarily property 1. The total length of the edges in U provides a lower bound L_2 on the

7.3 Lower Bounds

length of an optimum Hamiltonian cycle in undirected graph G. Hence, for undirected graphs the maximum branching algorithm can be replaced by the less complicated maximum spanning tree algorithm.

Associate a number u(x) with each vertex x. Let

$$a'(x,y) = a(x,y) + u(x) + u(y) \qquad (5)$$

for all (x,y). Under this transformation, the length of every Hamiltonian cycle is changed by the same amount. Thus, the shortest Hamiltonian cycle for the original edge lengths a(x,y) will also be the shortest Hamiltonian cycle for the new edge lengths a'(x,y). Held and Karp (1970, 1971) have devised methods to search for values u(x), x ∈ X, such that the set U produced above is a Hamiltonian cycle. These methods use nonlinear programming techniques and any further discussion of them would take us too far afield.

7.4 SOLUTION TECHNIQUES

Many solution techniques are available for the traveling salesman problem (Bellmore and Nemhauser, 1968; Garfinkel and Nemhauser. 1972; Held and Karp, 1970; 1971; Steckman, 1970). All of these techniques fall into one of two categories:

(a) Techniques that are certain to find an optimum solution but at worst require a prohibitive number of calculations

(b) Techniques that are not always certain to find an optimum solution but require a reasonable number of calculations.

A large number of the solution techniques rely heavily upon advanced results in integer linear programming, nonlinear programming, and dynamic programming, and a description of these techniques would take us too far afield. In this section, we shall confine ourselves to two solution techniques that require only results developed in earlier chapters. The first technique is the "branch and bound" technique which falls into category (a). The second technique is the "successive improvement" technique which falls into category (b).

Branch and Bound Solution Technique

Let $G = (X,A)$ denote the graph under consideration. As shown in Sec. 7.4, the minimum cost flow algorithm of Sec. 4.3 can be used to generate a lower bound L_1 on the length of a shortest Hamiltonian circuit in graph G. If the flow generated by the minimum cost flow algorithm corresponds to a circuit in graph G, then this circuit is an optimum Hamiltonian circuit, and the salesman problem has been solved. However, chances are that for any graph of realistic proportions, the resulting flow will correspond to several disjoint circuits. Select any one of these circuits and denote the vertices contained in this circuit by $X_c = \{x_1, x_2, \ldots, x_k\}$.

In an optimum solution, the salesman upon leaving vertex x_1 either travels to a vertex not in X_c or to a vertex in X_c. If the latter happens, then when the salesman leaves vertex x_2 he goes either to a vertex not in X_c or to a vertex in X_c, etc. Thus, an optimum solution must be present in at least one of the following graphs:

1. Graph G with all arcs (x_1,y), $y \in X_c$, deleted
2. Graph G with all arcs (x_1,y), $y \notin X_c$, deleted and arcs (x_2,z), $z \in X_c$, deleted
3. Graph G with all arcs (x_1,y), (x_2,y), $y \notin X_c$, deleted, and all arcs (x_3,z), $z \in X_c$, deleted
. . . .
. . . .
k. Graph G with all arcs (x_1,y), (x_2,y), $\ldots$, (x_{k-1},y), $y \notin X_c$, deleted, and all arcs (x_k,z), $z \in X_c$, deleted.

(Note that deleting an arc is equivalent to giving it an infinite length.) Call the above graphs G_1, G_2, $\ldots$, G_k, respectively. Since an arc length in a subscripted graph G_i is never less than the length of the corresponding arc in graph G, the length of an optimum Hamiltonian circuit in G_i cannot be less than the length of an optimum Hamiltonian circuit in G. Moreover, an optimum solution for graph G is also an optimum solution for at least one of the subscripted graphs G_1, G_2, $\ldots$, G_k.

7.4 Solution Techniques

Next, use the minimum cost flow algorithm to find a lower bound L_1 on the length of an optimum Hamiltonian circuit for each graph $G_1, G_2, \ldots, G_k$. Call these lower bounds $L_1(G_1)$, $L_1(G_2)$, $L_1(G_3)$, $\ldots$, $L_1(G_k)$, respectively. (The reader familiar with linear programming techniques may replace the minimum cost flow algorithm with the assignment algorithm.)

If the minimum cost flow in some subscripted graph G_c corresponds to a Hamiltonian circuit, then we need not give any further consideration to any subscripted graph G_j for which $L_1(G_j) \geq L_1(G_c)$ since $L_1(G_c)$ is an achievable lower bound. In this way, some of the subscripted graphs can be eliminated from further consideration.

For each remaining, noneliminated, subscripted graph G_i repeat the above procedure by replacing it with graphs $G_{i_1}, G_{i_2}, \ldots$. Calculate a lower bound $L_1(G_{i_j})$ on the length of an optimum Hamiltonian circuit for each graph G_{i_j} and proceed as above to eliminate as many double-subscripted graphs as possible from further consideration.

Ultimately, one graph in which a shortest Hamiltonian circuit has been found will eliminate all other graphs. This Hamiltonian circuit must be a shortest Hamiltonian circuit of the original graph G.

This technique is called the *branch and bound solution technique* because it branches from the original graph to other graphs and generates a bound on the optimal solution along each branch.

EXAMPLE. Let us use the branch and bound solution technique to find an optimum Hamiltonian circuit for graph G shown in Fig. 7.4. Since graph G is small, we can see immediately that the minimum cost flow algorithm applied to graph G (see Sec. 7.3 for details) would yield two circuits (a,b), (b,c), (c,a) and (f,e), (e,d), (d,f). The total length of these two circuits is 11. Thus, $L_1(G) = 11$. Using the vertices $X_c = \{a,b,c\}$ in the first circuit, we can generate three graphs G_a, G_b, G_c, shown in Fig. 7.4. Graph G_a is graph G without any arcs from a to b or c, i.e., without arc (a,b). Graph

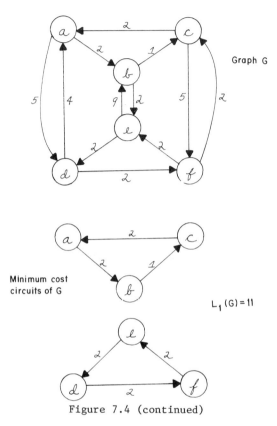

Figure 7.4 (continued)

G_b is graph G with all arcs from a to d, e, f deleted and all arcs from b to a and c deleted. Thus, G_b is G without arcs (a,d) and (b,c). Graph G_c is graph G with all arcs from a and b to d, e, f deleted and all arcs from c to a and b deleted. Thus, G_c is G without arcs (a,d), (b,e), and (c,a).

Since G_a, G_b, and G_c are small, we can see immediately that the minimum cost flow algorithm would yield the following circuits:

Graph	Circuit	Lower bound L_1
G_a	(a,d), (d,f), (f,c), (e,b), (b,c), (c,a)	$L_1(G_a) = 21$
G_b	(a,b), (b,e), (e,d), (d,f), (f,c), (c,a)	$L_1(G_b) = 12$
G_c	(a,b), (b,c), (c,f), (f,e), (e,d), (d,a)	$L_1(G_c) = 16$

7.4 Solution Techniques

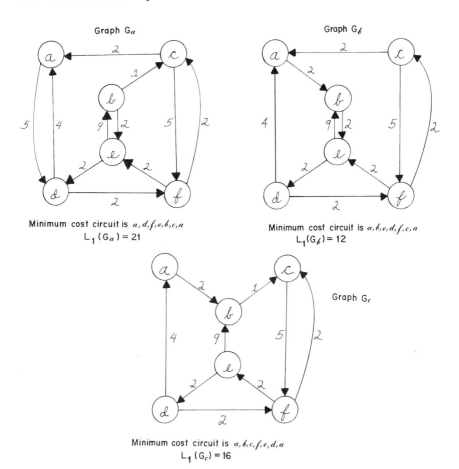

Example of Branch and Bound Solution Technique

The minimum cost flow in G_b generates a Hamiltonian circuit with length equal to 12. Consequently, graphs G_a and G_c can be eliminated from further consideration since their lower bounds both exceed $L_1(G_b) = 12$. Consequently, the Hamiltonian circuit (a,b), (b,e), (e,d), (d,f), (f,c), (c,a) is an optimum Hamiltonian circuit for the original graph G.

Successive Improvement Technique

The *successive improvement technique* for finding a short (and sometimes shortest) Hamiltonian circuit in graph G is as follows:

Start with any Hamiltonian circuit. Let $x_1, x_2, \ldots, x_n$ denote the order in which this circuit visits the vertices of graph G. For $i = 1, 2, \ldots, n - 1$ and $j = i + 1, i + 2, \ldots, n$ determine if switching vertices x_i and x_j in the preceding ordering generates a shorter Hamiltonian circuit. If so, make this switch in the order in which the vertices are visited. Repeat this process until no more switching is possible.

Switching vertices i and j in effect means replacing arcs (x_{i-1}, x_i), (x_i, x_{i+1}), (x_{j-1}, x_j), (x_j, x_{j+1}) by arcs (x_{i-1}, x_j), (x_j, x_{i+1}), (x_{j-1}, x_i), (x_i, x_{j+1}).

The switching process must eventually stop since (a) there are only a finite number of different ways to order the n vertices, and (b) no ordering is repeated since each switch generates a Hamiltonian circuit that is shorter than the preceeding Hamiltonian circuit.

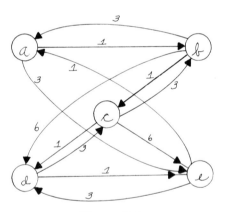

Figure 7.5

Example of Successive Improvement Technique

The final Hamiltonian circuit depends upon the initial Hamiltonian circuit. For example, consider graph G in Fig. 7.5. This graph contains five vertices; consequently, each Hamiltonian circuit must contain five arcs. Since all arcs are at least one unit long, no Hamiltonian circuit is less than five units long. The Hamiltonian circuit traversing the vertices in the order a, b, c, d, e is five units long and consequently must be an optimum Hamiltonian circuit.

7.4 Solution Techniques

Suppose that the successive improvement technique were initialized with the Hamiltonian circuit that traverses the vertices in the order a, e, d, c, b. The reader can verify that no two vertices in this sequence can be switched so that the resulting sequence corresponds to a Hamiltonian circuit in G. (For example, a and e cannot be switched since the sequence e, a, d, c, b does not represent a circuit because there is no arc from a to d.) Thus, if the successive improvement technique is initialized with a, e, d, c, b, it terminates with a, e, d, c, b. The corresponding Hamiltonian circuit has length equal to $3 + 3 + 3 + 3 + 3 = 15$, which clearly is not optimum.

On the other hand, suppose we started with a, b, d, c, e. The corresponding Hamiltonian circuit has length equal to $1 + 6 + 3 + 6 + 1 = 17$. If vertices c and d are switched, then the resulting sequence a, b, c, d, e corresponds to an optimum Hamiltonian circuit with length $1 + 1 + 1 + 1 + 1 = 5$.

Observe in the above example that the length of the initial Hamiltonian circuit is not necessarily a good indicator of the length of the final Hamiltonian circuit. When the technique was initialized with a Hamiltonian circuit of length 15, it terminated with a circuit with length 15. When the technique was initialized with a circuit with length 17, it terminated with a circuit with length 5.

In this example, it was easy to verify the optimality of a circuit. In general, we cannot be certain about the optimality of the terminal circuit produced by the successive improvement technique. In this situation, lower bounds L_1 and L_2 can be used to determine the maximum amount by which the terminal circuit exceeds optimality.

Methods for improving the successive improvement technique can be found in Steckman (1970) and Lin and Kernighan (1973).

EXERCISES

1. Show that the set of all optimum solutions to the salesman problems is the same for all starting vertices.

2. Suppose that the salesman must visit vertex b immediately after visiting vertex a. How can this be incorporated into the salesman problem?
3. Suppose that the salesman must visit vertex b after (not necessarily, immediately after) visiting vertex a. How can this be incorporated into the salesman problem?
4. Use the results of Sec. 7.3 to verify that the graphs shown in Fig. 7.6 and 7.7 possess Hamiltonian circuits.

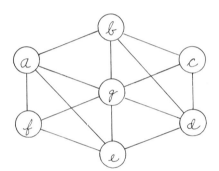

Figure 7.6

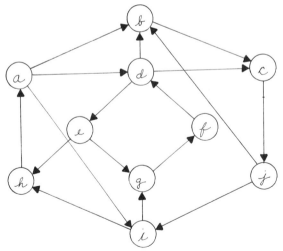

Figure 7.7

Exercises

5. For the graph in Fig. 7.3, we found that $L_1 < L_2$. Construct a graph for which $L_1 > L_2$. Construct a graph for which $L_1 = L_2$.
6. Use the branch and bound solution technique to find an optimum solution to the salesman problem for the graph shown in Fig. 7.8.

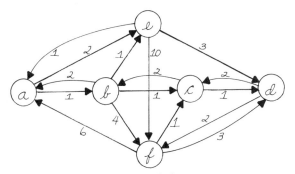

Figure 7.8

7. Use the successive improvement technique to find a short salesman route for the graph shown in Fig. 7.8. Start this technique with the Hamiltonian circuits that visits the vertices in the order d, c, b, a, e, f.
8. Suppose that two salesmen initialize the successive improvement technique with the same nonoptimum Hamiltonian circuit. The sequence of vertex switches used by the first salesman is different than that used by the second salesman. Show that their terminal circuits may be different.
9. In the past, all announcements for general circulation around the Operations Research Department of a large firm were mimeographed and one copy was given to each person in the department. In the interest of economy, the department has decided that only one copy of the announcement would be made and that it would be routed throughout the department along some predetermined route. The distance between the desks of each of the six people in the department are given in the following table. What is the best routing for an announcement? (The announcement must always return to its sender so that he can be certain that it was not lost somewhere along the route.)

From	To	Employee					
		1	2	3	4	5	6
Employee	1	0	9	8	7	6	10
	2		0	10	9	15	20
	3			0	5	15	25
	4				0	20	5
	5					0	20

(Distance in yards.)

10. Four machine operations must be performed on every job entering a machine shop. These operations may be performed in any sequence; however, the set-up time on a machine depends upon the previous operation performed on the job. The set up times are given in the following table. What is the best sequence of operations?

From	To	Machine			
		A	B	C	D
Machine	A	x	15	20	5
	B	30	x	30	15
	C	25	25	x	15
	D	20	35	10	x

REFERENCES

Bellmore, M., and G. L. Nemhauser, 1968. The Traveling Salesman Problem: A Survey, ORSA, vol. 16, pp. 538-558.

Berge, C., 1973. Graphs and Hypergraphs, (English translation by E. Minieka), North-Holland, Amsterdam, Chap. 10, pp. 186-227.

Christofides, N., 1972. Bounds for the Travelling-Salesman Problem, ORSA, vol. 20, no. 6, pp. 1044-1056.

Chvátal, V., 1972. On Hamilton Ideals, J. Combinatorial Theory, Series B, vol. 12, no. 2, pp. 163-168.

Garfinkel, R., and G. L. Namhauser, 1972. Integer Programming, John Wiley, Inc., New York, pp. 354-360.

References

Ghouila-Houri, A., 1960. Une condition suffisante d'existence d'un circuit hamiltonien, *Cahiers du Research Acad. Sci.*, vol. 251, pp. 494-497.

Held, M., and R. Karp, 1970. The Traveling-Salesman Problem and Minimum Spanning Trees, *ORSA*, vol. 18, pp. 1138-1162.

Held, M., and R. Karp, 1971. The Traveling-Salesman Problem and Minimum Spanning Trees, Part II, *Math. Programming*, vol. 1, no. 1, pp. 6-25.

Steckhan, H., 1970. A Theorem on Symmetric Traveling Salesman Problems, *ORSA*, vol. 18, pp. 1163-1167.

Lin, S., and B. W. Kernighan, 1973. An Effective Heuristic Algorithm for the Traveling-Salesman Problem, *ORSA*, vol. 21, no. 2, pp. 498-516.

Chapter 8

LOCATION PROBLEMS

8.1 INTRODUCTION

Location theory is concerned with the problem of selecting the best location in a specified region for a service facility such as a shopping center, fire station, factory, airport, warehouse, etc. The mathematical structure of a location problem depends upon the region available for the location and upon how we judge the quality of a location. Consequently, there are a large variety of different kinds of location problems, and the literature is filled with a variety of solution techniques.

In this chapter, we shall confine ourselves to location problems in which the region in which the facility is to be located is a graph, i.e., the facility must be located somewhere on an arc or at a vertex of a graph. Moreover, we shall confine ourselves to location problems that do not take us too far afield mathematically.

EXAMPLE 1. A county has decided to build a new fire station which must serve all six townships in the county. The fire station is to be located somewhere along one of the highways in the county so as to minimize the distance to the township farthest from the fire station.

If the highways of the county are depicted as the edges of a graph, this fire station location problem becomes the problem of locating the point on an edge with the property that the distance along the edges (highways) from this point to the farthest vertex (township) is as small as possible.

Consider the graph in Fig. 8.1. If vertex 3 were selected as the location, the most distant vertex from 3 is 6, which is 3 units away. Better still is vertex 2 which is at most two units away from any vertex. Even better is the midpoint of edge (2,5) which is 1 1/2 units away from vertices 1, 3, 4, and 6.

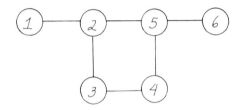

Figure 8.1

(All arc lengths equal 1.)

EXAMPLE 2. Suppose that the same county must locate a post office so that the total distance from the post office to all the townships is minimized. In this case, locating the post office at the midpoint of edge (2,5) would give a total distance of 1 1/2 + 1/2 + 1 1/2 + 1 1/2 + 1/2 + 1 1/2 = 7 units since the distance from the midpoint of edge (2,5) to vertices 1, 2, 3, 4, 5, 6 is respectively 1 1/2, 1/2, 1 1/2, 1 1/2, 1/2, 1 1/2 units.

Note that Examples 1 and 2 are essentially the same problem except that the criterion for judging locations is different. In Example 1, the maximum distance is minimized; in Example 2, the sum of the distances is minimized. A location selected according to the criterion, i.e., minimizing the maximum distance, is called a *center*. A location selected according to the latter criterion, i.e., minimizing the sum of the distances, is called a *median*. More rigorous definitions will be given later.

EXAMPLE 3. Suppose that the same county must locate a station for tow-trucks to rescue motorists who have become stranded somewhere on the county's highways. Suppose, also, that each potential location

8.1 Introduction

is judged according to the maximum distance that a towtruck must travel to rescue a motorist.

In this situation, the maximum distance to all points on all edges must be considered instead of the maximum distance to a vertex, as in Example 1.

EXAMPLE 4. Suppose that the same county must select a location for a telephone switching station somewhere in a town or along a highway. The switching station must be located so as to minimize the total length of all telephone lines that must be laid between itself and the six townships. To complicate matters, the townships have varying population sizes and require anywhere between one and five lines from themselves to the switching station.

Observe that if each township required only one line connecting itself with the switching station, then the problem posed in Example 4 would have a mathematical structure identical to the problem posed in Example 2. However, now certain townships must be considered more heavily and the location of the switching station must be influenced by the juxtaposition of the more populous townships.

Before making more rigorous definitions for the various types of locations to be considered, some definitions to describe the points on arcs and the various distances in a graph are needed.

Number the vertices of the graph G that is under consideration 1 through n. Consider any arc (i,j) whose length is given by $a(i,j) > 0$. Let the *f-point* of arc (i,j) denote the point on arc (i,j) that is $f\ a(i,j)$ units from vertex i and $(1 - f)\ a(i,j)$ units from vertex j for all f, $0 \leq f \leq 1$. Thus, the 1/4-point of arc (i,j) is 1/4 of the way along arc (i,j) from vertex i towards vertex j.

The 0-point of arc (i,j) is vertex i, and the 1-point of arc (i,j) is vertex j. Thus, the vertices may also be regarded as points. Points that are not vertices are called *interior points*. A point must be either an interior point or a vertex. As before, let X denote the set of all vertices. Let P denote the set of all points. Thus, P - X is the set of all interior points.

Let d(i,j) denote the length of a shortest path from vertex i to vertex j. Let D denote the n x n matrix whose i,j-th element is d(i,j). The elements in matrix D are called *vertex-vertex distances*. Recall from Sec. 3.2, that either the Floyd or Dantzig algorithms can be used to calculate matrix D.

Let d(f - (r,s),j) denote the length of a shortest path from the f-point on arc (r,s) to vertex j. This is called a *point vertex distance*. If arc (r,s) is undirected, i.e., allows travel in both directions, then this distance must be the smaller of the two following distances:

(a) The distance from the f-point to vertex r plus the distance from vertex r to vertex j

(b) The distance from the f-point to vertex s plus the distance from vertex s to vertex j.

Thus,

$$d(f - (r,s), j) = \min\{fa(r,s) + d(r,j), (1 - f)a(r,s) + d(s,j)\} \quad (1a)$$

If (r,s) is a directed arc, i.e., travel is allowed only from r to s, then the first term in the above minimization is eliminated, and

$$d(f - (r,s),j) = (1 - f)a(r,s) + d(s,j). \quad (1b)$$

Observe that only the arc lengths and the D matrix are needed to compute all the point-vertex distances.

When plotted as a function of f, the point-vertex distance for a given arc (r,s) and given vertex j must take one of the three forms shown in Fig. 8.2. Note that the slope of this piecewise linear curve is either +a(r,s) or -a(r,s), and the slope makes at most one change from +a(r,s) to -a(r,s).

Next, consider the shortest distance from vertex j to each point on arc (r,s). For some point on arc (r,s), this distance takes its maximum value. This maximum distance from vertex j to any point on arc (r,s) is denoted by d'(j,(r,s)) and is called a *vertex-arc distance*.

8.1 Introduction

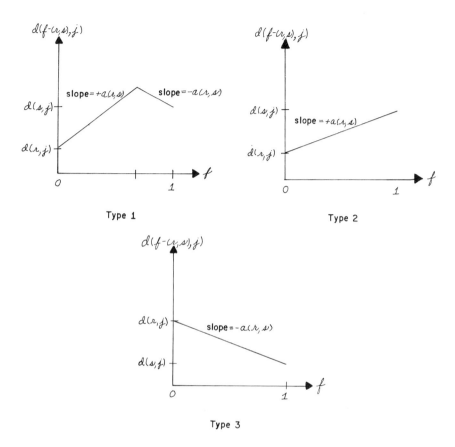

Figure 8.2
Plots of Point-Vertex Distances

If arc (r,s) is undirected, there are two ways to travel from vertex j to the f-point on (r,s): via vertex r or via vertex s. Naturally, we select the shorter of the two routes. If these two routes have unequal distances, then some neighboring points of the f-point of arc (r,s) are even further away from vertex j. For example, in Fig. 8.1, the 0.25-point of edge (3,4) is 1.25 units or 2.75 units away from vertex 2 depending if you travel via vertex 3 or via vertex 4. If f is increased from 0.25 to 0.26, then the shortest distance from vertex 2 to the 0.26-point of edge (3,4) is min{1,26,2.74} = 1.26. These two distances are equal at the most distant point. Observe that these two distances always sum to always sum to

$$d(j,r) + fa(r,s) + d(r,s) + (1 - f)a(r,s) = d(j,r) + d(j,s) + a(r,s)$$

Thus, it follows that

$$d'(j,(r,s)) = \frac{d(j,r) + d(j,s) + a(r,s)}{2} \qquad (2a)$$

If, on the other hand, arc (r,s) is directed, then a point on arc (r,s) can be reached only via vertex r. Consequently, the most distant points on (r,s) from any vertex are those points closest to vertex s, i.e., the f-points for which f approaches 1. In this case

$$d'(j,(r,s)) = d(j,r) + a(r,s) \qquad (2b)$$

Number the arcs in graph G one through m. Let D' denote the n x m matrix whose j,k-th elements is the vertex-arc distance from vertex j to arc k. Observe that the vertex-arc distance matrix D' can be computed from the vertex-vertex distance matrix D and the arc lengths by using equations (2a) and (2b).

Let $d'(f - (r,s), (t,u))$ denote the maximum distance from the f-point of arc (r,s) to the points on arc (t,u). This distance is called a *point-arc distance*.

If arc (r,s) is undirected and if $(r,s) \neq (t,u)$, then the route from the f-point on (r,s) to the most distant point on (t,u) must be either via vertex r or via vertex s. Thus, it follows that

$$d'(f - (r,s), (t,u)) = \min\{fa(r,s) + d'(r,(t,u)),$$
$$(1 - f)a(r,s) + d'(s,(t,u))\} \qquad (3a)$$

If arc (r,s) is directed and $(r,s) \neq (t,u)$, then the first term in the above minimization can be eliminated, and

$$d'(f - (r,s),(t,u)) = (1 - f)a(r,s) + d'(s,(t,u)) \qquad (3b)$$

If $(r,s) = (t,u)$, and if arc (r,s) is directed, then the most distant points on arc (r,s) from the f-point on (r,s) are the g-points where g approaches f from values less than f. Thus, in this case,

8.1 Introduction

$$d'(f - (r,s),(r,s)) = (1-f)a(r,s) + d(s,r) \quad (3c)$$

If $(r,s) = (t,u)$, and if arc (r,s) is undirected, then the maximum distance from the f-point on (r,s) to any g-point on (r,s) (where $g < f$) cannot exceed

$$A \equiv \min\{fa(r,s), 1/2[a(r,s) + d(s,r)]\}$$

The first term in this minimization accounts for routes from the f-point to the g-point restricted to arc (r,s). The second term in the minimization accounts for routes from the f-point on (r,s) to the g-point on (r,s) that traverse vertex s.

Similarly, the maximum distance from the f-point on (r,s) to any g-point on (r,s) (where $g > f$) cannot exceed

$$B = \min\{(1 - f)a(r,s), 1/2[a(r,s) + d(r,s)]\}$$

The first term in the preceding minimization accounts for routes from the f-point to the g-point restricted to arc (r,s). The second term in the preceding minimization accounts for routes from the f-point on (r,s) to the g-point on (r,s) that traverse vertex r.

Consequently, if arc (r,s) is undirected,

$$d'(f - (r,s), (r,s)) = \max\{A,B\}$$

or, equivalently,

$$d'(f - (r,s),(r,s)) = \max\{\min\{fa(r,s), 1/2[a(r,s) + d(s,r)]\}, \\ \min\{(1 - f)a(r,s), 1/2[a(r,s) + d(r,s)]\}\} \quad (3d)$$

When $d'(f - (r,s), (t,u))$ is plotted as a function of f for all $(r,s) \neq (t,u)$, the curve takes the same form as the point-vertex distances shown in Fig. 8.2, since equation (3a) has the same form as equation (1a) and equation (3b) has the same form as equation (1b). Only the constants are different, the equational forms are the same.

On the other hand, when $d'(f - (r,s),(r,s))$ is plotted as a function of f for any undirected arc (r,s), the curve takes the form shown in Fig. 8.3. This follows from equation (3d).

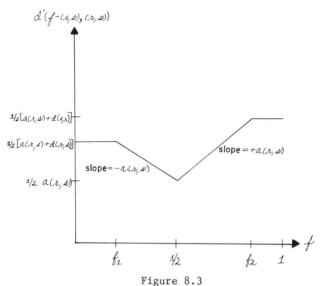

Figure 8.3
Plot of Point-Arc Distance
[d'(f - (r,s),(r,s))]

Symbol	Name	Equation
a(i,j)	Arc length	Given
d(i,j)	Vertex-vertex distance	VV Floyd or Dantzig algorithm
d(f - (r,s),j)	Point-vertex distance	PV (1a), (1b)
d'(j,(r,s))	Vertex-arc distance	VA (2a), (2b)
d'(f - (r,s),(t,u))	Point-arc distance	PA (3a),(3b),(3c),(3d)

Let

$$MVV(i) = \max_{j}\{d(i,j)\} \qquad (4)$$

denote the maximum distance of any vertex from vertex i.

Let

$$SVV(i) = \sum_{j} d(i,j) \qquad (5)$$

denote the total distance of all vertices from vertex i.

8.1 Introduction

Similarly, let

$$\text{MPV}(f - (r,s)) = \max_{j} \{d(f - (r,s), j)\} \quad (6)$$

denote the maximum distance of any vertex from the f-point on arc (r,s).

Similarly, let

$$\text{SPV}(f - (r,s)) = \sum_{j} d(f - (r,s), j) \quad (7)$$

denote the total distance of all vertices from the f-point on arc (r,s). In a similar manner we can define MVA(i), SVA(i), MPA(f - (r,s)), SPA(f - (r,s)) by taking maximums or sums over all arcs, rather than over all vertices as in equations (4)-(7).

With all these definitions for distances, maximums and sums, we are now ready to state rigorously the definitions of the various types of locations that we shall consider.

1. A *center* of graph G is any vertex x of graph G such that

$$\text{MVV}(x) = \min_{i} \{\text{MVV}(i)\} \quad (8)$$

Thus, a center is any vertex whose furthest vertex is as close as possible.

2. A *general center* of graph G is any vertex x of graph G such that

$$\text{MVA}(x) = \min_{i} \{\text{MVA}(i)\} \quad (9)$$

Thus, a general center is any vertex whose furthest point is as close as possible.

3. An *absolute center* of graph G is any f-point on any arc (r,s) of graph G such that

$$\text{MPV}(f - (r,s)) = \min_{f-(t,u) \in P} \{\text{MPV}(f - (t,u))\} \quad (10)$$

Thus, an absolute center is any point whose furthest vertex is as close as possible.

4. A *general absolute center* of graph G is any f-point on any arc (r,s) of G such that

$$\text{MPA}(f = (r,s)) = \min_{f-(t,u) \in P} \{\text{MPA}(f - (t,u))\} \qquad (11)$$

Thus, a general absolute center is any point whose furthest point is as close as possible.

Analogous to each of the four kinds of locations defined here, we can define a (5) *median*, (6) *general median*, (7) *absolute median*, and (8) a *general absolute median*. The definitions are analogous to the preceding definitions except that everywhere the maximization operation [that is, MVV(i), MVA(i), MPV(f - (t,u)), MPA(f - (t,u))] is replaced by the summation operation [that is, SVV(i), SVA(i), SPV(f - (t,u)), SPA(f - (t,u))].

Observe that Example 1 calls for an absolute center; Example 2 calls for an absolute median; Example 3 calls for a general absolute center, and Example 4 calls for a "weighted absolute median", which is discussed in Sec. 8.4.

In Sec. 8.2, methods are developed for finding the four kinds of centers defined above. In Sec. 8.3, methods are developed for finding the four kinds of medians defined in the preceding discussion.

8.2 CENTER PROBLEMS

This section presents a method for finding each of the four types of centers described in Sec. 8.1.

Center

Recall that a center is any vertex x with the smallest possible value of MVV(x), i.e., a center is any vertex x with the property that the most distant vertex from x is as close as possible.

The Floyd algorithm or the Dantzig algorithm of Sec. 3.2 can be used to calculate the vertex-vertex distance matrix D whose i,j-th element is d(i,j) the length of a shortest path from vertex i to vertex j. The maximum distance MVV(i) of any vertex from vertex i is the largest entry in the i-th row of matrix D. A center is any vertex x with the smallest possible value of MVV(x), i.e., a center is any vertex whose row in the D matrix has the smallest maximum entry.

8.2 Center Problems

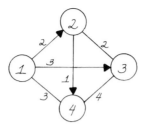

Arc	Order
1.	(1,2)
2.	(1,3)
3.	(1,4)
4.	(2,4)
5.	(2,3)
6.	(3,4)

Figure 8.4

Computational Example

EXAMPLE 1. Let us find a center for the graph shown in Fig. 8.4. It is left to the reader to verify by using either the Floyd or Dantzig algorithm that

$$D = \begin{bmatrix} 0 & 2 & 3 & 3 \\ 4 & 0 & 2 & 1 \\ 6 & 2 & 0 & 3 \\ 3 & 5 & 4 & 0 \end{bmatrix}$$

Thus,

$MVV(1) = \max\{0,2,3,3\} = 3$

$MVV(2) = \max\{4,0,2,1\} = 4$

$MVV(3) = \max\{6,2,0,3\} = 6$

$MVV(4) = \max\{3,5,4,0\} = 5$

Thus, $\min_i MVV(i) = \min\{3,4,6,5\} = 3 = MVV(1)$. Consequently, vertex 1 is a center of this graph. The farthest vertex from vertex 1 is three units away. No other vertex can do better than 3 units.

GENERAL CENTER

Recall that a general center is any vertex x with the smallest possible value of MVA(x), i.e. a general center is any vertex such that the most distant point from vertex x is as close as possible.

A general center can be found by finding the row of D' with the smallest maximum entry. This row corresponds to a vertex that is a general center. This follows since MVA(i) equals the largest entry in the i-th row of the vertex-arc distance matrix D'.

EXAMPLE 2. Let us find a general center for the graph in Fig. 8.4. The arcs of this graph are numbered 1 through 6 in Fig. 8.4. Using the vertex-vertex distance matrix D given in the preceeding example and the arc lengths given in Fig. 8.4, we can use equations (2) to calculate

$$D' = \begin{bmatrix} 2 & 3 & 3 & 3 & 3\ 1/2 & 5 & - \\ 6 & 7 & 4 & 1 & 2 & & 3\ 1/2 \\ 8 & 9 & 6 & 3 & 2 & & 3\ 1/2 \\ 5 & 6 & 3 & 6 & 5\ 1/2 & 4 & - \end{bmatrix}$$

For example, from equation (2a),

$$d'(1,(3,4)) = 1/2[d(1,3) + d(1,4) + a(3,4)]$$
$$= 1/2(3 + 3 + 4) = 5$$

From equation (2b),

$$d'(1,(2,4)) = d(1,2) + a(2,4) = 2 + 1 = 3$$

Thus,

$$MVA(1) = \max\{2,3,3,3,3\ 1/2,5\} = 5$$
$$MVA(2) = \max\{6,7,4,1,2,3\ 1/2\} = 7$$
$$MVA(3) = \max\{8,9,6,3,2,3\ 1/2\} = 9$$
$$MVA(4) = \max\{5,6,3,6,5\ 1/2,4\} = 6$$

Thus, min MVA(i) = min{5,7,9,6} = 5 = MVA(1). Hence, vertex 1 is a general center of the graph. The most distant point from vertex 1 is 5 units away from vertex 1 and lies on arc (3,4).

ABSOLUTE CENTER

Recall that an absolute center is any point whose most distant vertex is as close as possible. To find an absolute center, we must find the point f - (r,s) such that $MPV(f - (r,s)) = \min_{f-(t,u) \in P} MPV(f - (t,u))$.

8.2 Center Problems

The absolute center problem is more difficult than the center problem or general center problem because all points, not only the vertices, must be considered.

First, an observation: *No interior point of a directed arc can be an absolute center.* Since all travel on a directed arc is in one direction, it follows that the terminal vertex of a directed arc is closer each vertex in the graph than is any interior point of the directed arc. Consequently, we need consider only vertices and interior points of undirected arcs in our search for an absolute center.

Consider any undirected arc (r,s). The distance $d(f - (r,s),j)$ from the f-point on (r,s) to vertex j is given by equation (1a) and plotted in Fig. 8.2. This distance is easy to plot as a function of f since its plot is a piecewise-linear curve with at most two pieces.

Plot $d(f - (r,s),j)$ for all f, $0 \leq f \leq 1$, for all vertices j. The uppermost portion of all these plots represents $\max_j d(f - (r,s),j)$. The value f^* of f at which this uppermost portion of all these plots takes its minimum value is the best candidate for absolute center on edge (r,s).

The best candidate on each undirected arc must be located by this method. The absolute center is any candidate $f^* - (r,s)^*$ with the minimum distance to its furthest vertex, i.e.

$$\max_j \{d(f^* - (r,s)^*,j)\} = \min_{(t,u)} \{\max_j d(f^* - (t,u),j)\}$$

To summarize, a candidate for absolute center is found by selecting the point on each undirected arc whose most distant vertex is as close as possible. The candidate with smallest distance between itself and its most distant vertex is selected as an absolute center. The selection of the candidate on each edge requires the plotting of all point-vertex distances as a function of the points on the edge. This is relatively uncomplicated since these distance functions are piecewise linear curves with at most two pieces. Unfortunately, there seems to be no way to avoid the plotting of the point-vertex distances. This method is due to Hakimi (1964).

EXAMPLE 3. Let us find an absolute center for the graph in Fig. 8.4. We know that all absolute centers (there may be ties, and consequently, more than one absolute center) must be either vertices or interior points of undirected arcs. The best vertex candidate for absolute center would be the vertex selected as center. In the example for calculating the center of this graph, we found that vertex 1 was the center and all vertices were within 3 units of vertex 1. Thus, vertex 1 is the best vertex candidate with a range of 3 units.

It remains to examine the interiors of the three undirected arcs (1,4), (2,3), and (3.4).

First, let us examine edge (3,4). From equation (1a),

$$d(f - (3,4),1) = \min\{fa(3,4) + d(3,1), (1-f)a(3,4) + d(4,1)\}$$
$$= \min\{4f + 6, 4(1-f) + 3\}$$
$$= \begin{cases} 4f + 6 & \text{For } f \leq \frac{1}{8} \\ 7 - 4F & \text{For } f \geq \frac{1}{8} \end{cases}$$

$$d(f - (3,4),2) = \min\{fa(3,4) + d(3,2), (1-f)a(3,4) + d(4,2)\}$$
$$= \min\{4f + 2, 4(1-f) + 5\}$$
$$= \begin{cases} 4f + 2 & \text{For } f \leq \frac{7}{8} \\ 9 - 4f & \text{For } f \geq \frac{7}{8} \end{cases}$$

$$d(f - (3,4),3) = \min\{fa(3,4) + d(3,3), (1-f)a(3,4) + d(4,3)\}$$
$$= \min\{4f, 4(1-f) + 4\}$$
$$= 4f \quad (\text{for all } f, 0 \leq f \leq 1)$$

$$d(f - (3,4),4) = \min\{fa(3,4) + d(3,4), (1-f)a(3,4) + d(4,4)\}$$
$$= \min\{4f + 3, 4(1-f) + 0\}$$
$$= \begin{cases} 4f + 3 & \text{For } f \leq \frac{1}{8} \\ 4 - 4f & \text{For } f \geq \frac{1}{8} \end{cases}$$

These point-vertex distances are plotted in Fig. 8.5. The lowest value taken by the uppermost portion of these curves occurs when $d(f - (3,4), 1) = d(f - (3,4), 2)$. Thus,

$$7 - 4f^* = 4f^* + 2, \quad f^* = 5/8$$

8.2 Center Problems

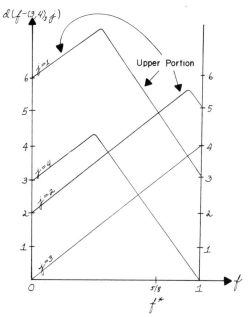

Figure 8.5
Plot of Point-Vertex Distances d(f - (3,4), j)

d(5/8 - (3,4), 1) = d(5/8 - (3,4), 2) = 4 1/2

Consequently, the 5/8-point is the best candidate for absolute center on edge (3,4), and no vertex is more than 4 1/2 units away from the 5/8-point of edge (3,4).

Figure 8.6 shows the same result for edge (1,4). Note that the best candidate for absolute center on edge (1,4) is the 0-point which is vertex 1. From before, we know that every vertex is within 3 units of vertex 1.

Figure 8.7 shows the same result for edge (2,3). Note that the best candidate for absolute center on edge (2,3) is the 0-point which is vertex 2. From before, we know that every vertex is within 4 units of vertex 2.

Consequently, the best interior point candidate is the 5/8-point on edge (3,4) with a maximum distance of 4 1/2 units. The best vertex dandidate is vertex 1 with a maximum distance of 3 units. Hence, vertex 1 is the absolute center of this graph.

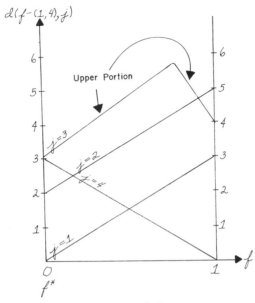

Figure 8.6

Plot of Point-Vertex Distances d(f - (1,4), j)

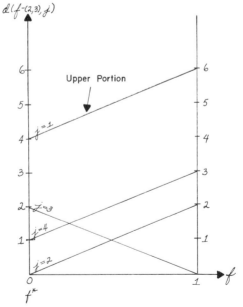

Figure 8.7

Plot of Point-Vertex Distances d(f - (2,3), j)

8.2 Center Problems

General Absolute Center

Recall that a general absolute center is any point x such that the furthest point from point x is as close as possible. To find a general absolute center we must find a point f - (r,s) such that

$$MPA(f - (r,s)) = \min_{f - (t,u) \in P} \{MPA(f - (t,u))\}$$

No interior point of a directed arc can be a general absolute center. Since all travel on a directed arc is in one direction, the terminal vertex of a directed arc is a better candidate for general absolute center than is any interior point of this arc since the terminal vertex is closer to every arc in the graph. Consequently, we need only consider vertices and the interior points of undirected arcs in our search for a general absolute center.

Observe that the problem of finding a general absolute center is identical to the problem for finding an absolute center, except we must now consider point-arc distances in place of point-vertex distances. As noted in Sec. 8.1, all the point-arc distance functions have the same form as the point-vertex distance functions, except the point-arc distance function d'(f - (r,s), (r,s)). The former have the form shown in Fig. 8.2; the latter has the form shown in Fig. 8.3.

In most realistic problems, the most distant point from the f-point on edge (r,s) will not lie on edge (r,s). In this case, we can simply omit from further consideration the point-arc distance function d'(f - (r,s), (r,s)). The problem of finding a general absolute center can now be solved by the technique used for finding an absolute center given above. The only difference is that the point-arc distance functions must replace the point-vertex distance functions. As there are more arcs than vertices, more plotting is required to find the general absolute center.

However, if there is a possibility that the most distant point from the f-point on edge (r,s) lies also on edge (r,s), then the plot of the point-arc distance function d'(f - (r,s), (r,s)) must be included in the calculations for the best candidate on edge (r,s).

Equation (3d) can be used to construct this plot. Happily, this plot is also piecewise linear with at most four pieces. See Fig. 8.3.

In summary, the technique for finding a general absolute center is the same as the technique for finding an absolute center except that the point-vertex distances are replaced by the point-arc distances.

8.3 MEDIAN PROBLEMS

This section presents methods for finding the four types of medians described in Sec. 8.1.

Median

Recall that a median is any vertex x with the smallest possible total distance from x to all other vertices. Thus, a median is any vertex x such that

$$SVV(x) = \min_{i}\{SVV(i)\}$$

The sum of the entries in the i-th row of the vertex-vertex distance matrix D equals the sum of the distances from vertex i to all other vertices, that is, SSV(i). Hence, a median corresponds to any row of D with the smallest sum.

EXAMPLE 1. Find a median of the graph in Fig. 8.4. From the previous examples, we know that the vertex-vertex distance matrix for this graph is

$$D = \begin{bmatrix} 0 & 2 & 3 & 3 \\ 4 & 0 & 2 & 1 \\ 6 & 2 & 0 & 3 \\ 3 & 5 & 4 & 0 \end{bmatrix}$$

Thus,

$$SVV(1) = 0 + 2 + 3 + 3 = 8$$
$$SVV(2) = 4 + 0 + 2 + 1 = 7$$
$$SVV(3) = 6 + 2 + 0 + 3 = 11$$
$$SVV(4) = 3 + 5 + 4 + 0 = 12$$

8.3 Median Problems

Hence, $\min_{i}\{SVV(i)\} = \min\{8,7,11,12\} = 7 = SVV(2)$, and vertex 2 is the median of this graph. The total distance from vertex 2 to all other vertices is 7 units.

General Median

A general median is any vertex x with the smallest total distance to each arc, where the distance from a vertex to an arc is taken to be the maximum distance from the vertex to the points on the arc. Thus, a general median is any vertex x such that

$$SVA(x) = \min_{i}\{SVA(i)\}$$

The sum of the entries in the i-th row of the vertex-arc distance matrix D' equals the sum of the distances from vertex i to all arcs, that is, SVA(i). Hence, a median corresponds to any row of D' with the smallest sum.

EXAMPLE 2. Find a general median for the graph in Fig. 8.4. From previous examples, we know that the vertex-arc distance matrix for this graph is

$$D' = \begin{bmatrix} 2 & 3 & 3 & 3 & 3\ 1/2 & 5 & - \\ 6 & 7 & 4 & 1 & 2 & & 3\ 1/2 \\ 8 & 9 & 6 & 3 & 2 & & 3\ 1/2 \\ 5 & 6 & 3 & 6 & 5\ 1/2 & 4 & - \end{bmatrix}$$

Thus,

$SVA(1) = 2 + 3 + 3 + 3 + 3\ 1/2 + 5 = 19$
$SVA(2) = 6 + 7 + 4 + 1 + 2 + 3\ 1/2 = 23\ 1/2$
$SVA(3) = 8 + 9 + 6 + 3 + 2 + 3\ 1/2 = 31\ 1/2$
$SVA(4) = 5 + 6 + 3 + 6 + 5\ 1/2 + 4 = 29\ 1/2$

Hence, $\min_{i}\{SVA(i)\} = \min\{19\ 1/2, 23\ 1/2, 31\ 1/2, 29\ 1/2\} = 19\ 1/2 = SVA(1)$. Thus, vertex 1 is the general median of this graph. The total distance from vertex 1 to all arcs is 19 1/2 units.

Absolute Median

An absolute median is any point whose total distance to all vertices is as small as possible.

THEOREM 8.1 There is always a vertex that is an absolute median.

Proof: Consider each point-vertex distance $d(f - (r,s), j)$ as a function of f. When plotted, this function takes the forms shown in Fig. 8.2. Observe that each function of this form has the property that if any two points on the curve are connected by a straight line see Fig. 8.8, then this straight line always lies on or below the curve. Any function with this property is called a *concave function*. Moreover, the minimum value of a concave function always occurs at one of its boundary points, i.e., either at $f = 0$ or at $f = 1$.

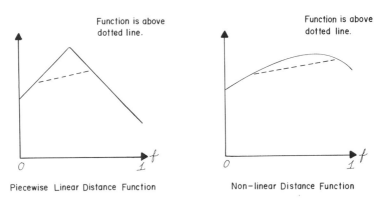

Piecewise Linear Distance Function   Non-linear Distance Function

Figure 8.8
Concave Distance Functions

Next, consider $SPV(f - (r,s)) = \sum_j d(f - (r,s), j)$, as a function of f. Since this function is the sum of concave functions, it too must be concave. Thus, $SPV(f - (r,s))$ is minimized at either $f = 0$ or at $f = 1$. Consequently, no interior point of edge (r,s) is a better candidate for absolute median than one of its end vertices. Q.E.D.

Observe that the proof of Theorem 8.1 remains valid not only for the point-vertex distances defined by equations (1) but also for any concave point-vertex distance function of f.

8.3 Median Problems

As a consequence of Theorem 8.1, we need consider only the vertices in our search for an absolute median. Thus, *any median is also an absolute median*, and no new solution techniques are needed.

General Absolute Median

A general absolute median is any point with the property that the total distance from it to all arcs is as small as possible. Again, the distance from a point to an arc is taken as the maximum distance from the point to all points on the arc. Thus, a general absolute median is any point $f - (r,s)$ such that

$$SPA(f - (r,s)) = \min_{f-(t,u) \in P} \{SPA(f - (t,u))\}$$

Theorem 8.1 stated that there always is a vertex that is an absolute median. The proof of Theorem 8.1 rested upon the fact that all the point-vertex distance functions were concave. If all the point-arc distance functions were also concave, then an analogous theorem could be proved for the general absolute median. Unfortunately, this is not the case since the point-arc distance function $d'(f - (r,s), (r,s))$ is not concave as can be seen in Fig. 8.3. Otherwise, all point-arc distance functions are concave as seen from Fig. 8.2. Consequently, it is possible all general absolute medians are interior points. (An example of this is given later.) It is possible to eliminate from consideration the interiors of some arcs. First, observe that *no interior* point of a directed arc can be a general absolute median. This follows since the terminal vertex of a directed arc is a better candidate for general absolute median than any interior point of this directed arc. Moreover,

THEOREM 8.2 No interior point of undirected arc (r,s) is a general absolute median if

$$|SVA(r) - SVA(s)| > 1/2[d(r,s) + d(s,r)] \tag{12}$$

Proof: The total distance from vertex r to all arcs is

$$SVA(r) = d'(r,(r,s)) + \sum_{(x,y):(x,y) \neq (r,s)} d'(r,(x,y)) \tag{13}$$

Likewise,

$$SVA(s) = d'(s,(r,s)) + \sum_{(x,y):(x,y) \neq (r,s)} d'(s,(x,y)) \quad (14)$$

Without loss of generality, suppose that $SVA(r) \leq SVA(s)$.

Since $d'(f - (r,s), (t,u))$ is concave for all $(r,s) \neq (t,u)$, it follows that $\sum_{(x,y):(x,y) \neq (r,s)} d'(f - (r,s),(x,y))$ lies above the straight line connecting its endpoints which are

$$\sum_{(x,y):(x,y) \neq (r,s)} d'(0 - (r,s), (x,y)) =$$

$$= \sum_{(x,y):(x,y) \neq (r,s)} d'(r,(x,y))$$

and

$$\sum_{(x,y):(x,y) \neq (r,s)} d'(1 - (r,s), (x,y))$$

$$= \sum_{(x,y):(x,y) \neq (r,s)} d'(s,(x,y))$$

Thus, when $f = 1/2$, it follows that

$$\sum_{(x,y):(x,y) \neq (r,s)} d'(1/2 - (r,s),(x,y))$$

$$\geq 1/2 \sum_{(x,y):(x,y) \neq (r,s)} [d'(s,(x,y)) + d'(r,(x,y))]$$

Hence, as f increases from zero to $1/2$, the quantity $\sum_{(x,y):(x,y)\neq(r,s)} d'(f - (r,s), (x,y))$ increases by at least $1/2 \sum_{(x,y):(x,y) \neq (r,s)} [d'(s,(x,y)) - d'(r,(x,y))]$.

Next, let us examine $d'(f - (r,s), (r,s))$. This function takes its minimum value at $f = 1/2$, which is $d'(1/2 - (r,s), (r,s))$. See equation (3d) and Fig. 8.3. As f goes from zero to $1/2$, $d'(f - (r,s), (r,s))$ experiences a decrease equal to $d'(r,(r,s)) - 1/2a(r,s)$.

If $\sum_{(x,y)} d'(f - (r,s), (x,y)) < SVA(r)$ for some value of f, $0 < f < 1$, then it is necessary that the maximum decrease of $d'(f - (r,s), (r,s))$ at $f = 1/2$ equal or exceed the minimum increase of $\sum_{(x,y):(x,y) \neq (r,s)} d'(f - (r,s), (x,y))$ at $f = 1/2$. In other words,

8.3 Median Problems

if an interior point of edge (r,s) is to be a better candidate for general absolute median than vertex r, it is necessary that

$$d'(r,(r,s)) - 1/2\, a(r,s) \geq 1/2 \sum_{(x,y):(x,y) \neq (r,s)} [d'(s,(x,y)) - d'(r,(x,y))] \quad (15)$$

Equivalently, it is necessary that

$$d'(s,(r,s)) + d'(r,(r,s)) - a(r,s) \geq SVA(s) - SVA(r) \quad (16)$$

Substituting equation (2a) into inequality (16) yields

$$1/2(d(s,r) + a(r,s)) + 1/2(d(r,s) + a(r,s)) - a(r,s)$$
$$\geq SVA(s) - SVA(r) \quad (17)$$

Simplifying inequality (17) yields

$$1/2(d(r,s) + d(s,r)) \geq SVA(s) - SVA(r) \quad (18)$$

If we had initially assumed that $SVA(s) \leq SVA(r)$, then inequality (18) would become

$$1/2(d(r,s) + d(s,r)) \geq SVA(r) - SVA(s) \quad (19)$$

Combining inequalities (18) and (19) yields inequality (12). Q.E.D.

Theorem 8.2 is useful because it provides an easy way to eliminate edges from further consideration in our search for a general absolute median. To check condition (12), only the vertex-vertex distance matrix D and the vertex-arc distance matrix D' are needed.

If not all edges are eliminated by Theorem 2, are some further eliminations possible? Yes:

LEMMA 8.1 For any interior point f on any edge (r,s),

$$SPA(f - (r,s)) \geq SVA(r) - 1/2\, d(r,s) \quad (20)$$

and

$$SPA(f - (r,s)) \geq SVA(s) - 1/2\, d(s,r) \quad (21)$$

Proof: From the proof of Theorem 8.2, we know that as f increases from zero to $1/2$, the distance $d'(f - (r,s), (r,s))$ decreases by $d'(r,(r,s)) - 1/2\, a(r,s)$, which by equation (2a) equals $1/2\, d(r,s)$. Thus, condition (20) follows.

If f is decreased from 1 to 1/2, then d'(f - (r,s), (r,s)) decreases by d'(s,(r,s)) - 1/2 a(r,s) = 1/2 d(s,r), and condition (21) follows. Q.E.D.

Lemma 8.1 can be used to generate a lower bound on the total distance for every interior point on any edge that was not eliminated by Theorem 8.2. Each of these edge lower bounds can be compared to the least total distance from a vertex, namely $\min_i\{SVA(i)\}$. If the lower found for an edge is greater the least total distance from a vertex, this edge can be eliminated.

Each remaining, noneliminated edge (r,s) must then be examined completely by evaluating SPA(f - (r,s)) for all f. Hopefully, the best candidate for general absolute median on the interior of the examined edge (r,s) will have a total distance that will be less then the lower bound of some nonexamined edges. In this case, these nonexamined edges can also be eliminated.

Ultimately, all edges must be either eliminated or completely examined. A general absolute median is selected from the set of vertices and interior point candidates.

EXAMPLE 3. Find a general absolute median for the graph in Fig. 8.1. Using either the Floyd algorithm or the Dantzig algorithm of Sec. 3.2, the vertex-vertex distance matrix D is found to be

$$D = \begin{bmatrix} 0 & 1 & 2 & 3 & 2 & 3 \\ 1 & 0 & 1 & 2 & 1 & 2 \\ 2 & 1 & 0 & 1 & 2 & 3 \\ 3 & 2 & 1 & 0 & 1 & 2 \\ 2 & 1 & 2 & 1 & 0 & 1 \\ 3 & 2 & 3 & 2 & 1 & 0 \end{bmatrix}$$

Next, order the edges as follows:

1. (1,2)
2. (2,3)
3. (3,4)
4. (4,5)
5. (5,6)
6. (2,5)

8.3 Median Problems

From equation (2a), the vertex-arc distance matrix d' can be calculated yielding

$$D' = \begin{bmatrix} 1 & 2 & 3 & 3 & 3 & 2 \\ 1 & 1 & 2 & 2 & 2 & 1 \\ 2 & 1 & 1 & 2 & 3 & 2 \\ 3 & 2 & 1 & 1 & 2 & 2 \\ 2 & 2 & 2 & 1 & 1 & 1 \\ 3 & 3 & 3 & 2 & 1 & 2 \end{bmatrix}$$

Thus,

SVA(1) = 1 + 2 + 3 + 3 + 3 + 2 = 14
SVA(2) = 1 + 1 + 2 + 2 + 2 + 1 = 9
SVA(3) = 2 + 1 + 1 + 2 + 3 + 2 = 11
SVA(4) = 3 + 2 + 1 + 1 + 2 + 2 = 11
SVA(5) = 2 + 2 + 2 + 1 + 1 + 1 = 9
SVA(6) = 3 + 3 + 3 + 2 + 1 + 2 = 14

Consequently, vertices 2 and 5 are the best vertex candidates for general absolute median since each of these vertices has a total distance to all arcs equal to 9 units.

Next, try to eliminate the interiors of some of the edges by applying condition (12) of Theorem 8.2. Observe that in this graph all arcs are undirected, and consequently, the right side of condition (12) becomes $d(r,s)$.

1. Edge (1,2) is eliminated because

   $|SVA(1) - SVA(2)| = |14 - 9| = 5 > 1 = d(1,2)$

2. Edge (2,3) is eliminated because

   $|SVA(2) - SVA(3)| = |9 - 11| = 2 > 1 = d(2,3)$

3. Edge (3,4) is not eliminated because

   $|SVA(3) - SVA(4)| = |11 - 11| = 0 < 1 = d(3,4)$

4. Edge (4,5) is eliminated because

   $|SVA(4) - SVA(5)| = |11 - 9| = 2 > 1 = d(4,5)$

5. Edge (5,6) is eliminated because

$$|SVA(5) - SVA(6)| = |9 - 14| = 5 > 1 = d(5,6)$$

6. Edge (2,5) is not eliminated because

$$|SVA(2) - SVA(5)| = |0 - 0| = 0 < 1 = d(2,5)$$

Thus, only edges (3,4) and (2,5) remain under consideration. Next, let us apply conditions (20) and (21) of Lemma 8.1 to see if any further edge eliminations are possible.

1. Edge (3,4) can be eliminated by condition (20) because

$$SPA(f - (3,4)) \geq SVA(3) - 1/2\ d(3,4) = 11 - 1/2 = 10\ 1/2$$

which is greater than the 9 units achieved by selecting a vertex as general absolute median.

2. Edge (2,5) cannot be eliminated by conditions (20) or (21) because

$$SPA(f - (2,5)) \geq SPA(2) - 1/2\ d(2,5) = 9 - 1/2 < 9$$

and

$$SPA(f - (2,5)) \geq SPA(5) - 1/2\ d(5,2) = 9 - 1/2 < 9$$

Only edge (2,5) remains under consideration. Equations (3a) and (3d) can be used to generate the point-arc distances for edge (2,5):

$$d'(f - (2,5),(1,2)) = 1 + f$$
$$d'(f - (2,5),(2,3)) = 1 + f$$
$$d'(f - (2,5),(3,4)) = 2$$
$$d'(f - (2,5),(4,5)) = 1 + (1 - f)$$
$$d'(f - (2,5),(5,6)) = 1 + (1 - f)$$
$$d'(f - (2,5),(2,5)) = \max\{f, (1 - f)\}$$

Adding these point-arc distances yields

$$SPA(f - (2,5)) = (1 + f) + (1 + f) + 2 + (2 - f) + (2 - f) + \max\{f, (1 - f)\}$$
$$= 8 + \max\{f, (1 - f)\}$$

Since $\min\limits_{0 \leq f \leq 1} \max \{f, (1 - f)\}$ occurs at $f = 1/2$, the best candidate for general absolute median on edge (2,5) is the 1/2-point, since

8.3 Median Problems 315

SPA(1/2 - (2,5)) = 8 1/2. Since 9 is the best possible total distance of a vertex, we conclude that the 1/2-point of edge (2,5) is the general absolute median of this graph with a total distance to all arcs equal to 8 1/2 units.

8.4 EXTENSIONS

Weighted Locations

In the preceeding sections, every vertex carried the same weight in the selection of a location. Every arc carried the same weight in the selection of a location. However, as shown in Sec. 8.1, Example 4, there are practical reasons for associating a different weight to each vertex and multiplying the distance to a vertex by the vertex's weight. Similarly, there are situations in which the distance to an arc should be multiplied by the arc's weight. For example, if the arcs represent highway segments that must be served from a central emergency station, each segment should be weighted with regard to the amount of traffic it carries.

The methods of Secs. 8.2 and 8.3 can be easily extended to find the various weighted locations, i.e., weighted center, weighted general center, weighted absolute center, etc. Merely, multiply each distance (i.e., each vertex-vertex distance, each vertex-arc distance, each point-vertex distance, each point-arc distance) by the weight associated with its destination vertex or destination arc. If these weighted distances replace the original non-weighted distances, the various methods of Secs. 8.2 and 8.3 will generate the corresponding weighted locations.

Multicenters and Multimedians

Secs. 8.2 and 8.3 were concerned with the problem of selecting exactly one location to serve as a center or median. Suppose, instaed, that we are allowed to select several facility locations. Each vertex (or arc) would then be associated with the location closest to it.

These multilocation problems are very complicated in that they consist of two stages: (a) the partition of the vertices (or arcs), and (b) the selection of the best location to serve all members of each subset of vertices (or arcs). Unfortunately, the techniques (Christofides and Viola, 1971; Marsten, 1972; Minieka, 1970; ReVelle and Swain, 1970; Toregas, et al., 1971) that are available for these problems ultimately rely upon integer programming for a final solution. Any detailed discussion of these techniques would take us too far afield and would best be treated in a text on integer programming.

However, one result concerning multi-absolute medians is worth noting (Hakimi, 1965; Goldman, 1969). Suppose we are searching for a set of p locations, p > 1, such that each vertex is associated with the location closest to it and the total distance from each location to the vertices associated with it is minimized. Such a set of points is called a p-absolute median.

THEOREM 8.3. There is a p-absolute median that consists entirely of vertices.

Proof: Theorem 8.1 proved this result for p = 1. For p > 1, the vertices are partitioned in p sets such that each set of vertices is served by the same median. Since *any* sum of point-vertex distance functions is a concave function, we know that each vertex set is optimally served by a median that is a vertex. Q.E.D.

EXERCISES

1. The four towns a, b, c, d in our county are connected by roads as shown in Fig. 8.9. Construct the D and D' matrices for this graph. Next, calculate a
    (a) Center
    (b) Absolute center
    (c) General center
    (d) General absolute center
    (e) Median
    (f) Absolute mean

Exercises

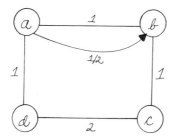

Figure 8.9

  (g) General median
  (h) General absolute median

2. Suppose that an additional road is built connecting cities b and d in Exercise 1. This road has length equal to 1 1/2. Repeat Exercise 1. (Assume the new road is a two-way road.)
3. Suppose that edge can be traveled in both directions, but that due to wind direction, the distance from x to y is 5 and the distance from y to x is 7. How can an edge such as (x,y) be incorporated into our models for finding centers and medians. (Note that you cannot replace this edge with two arcs (x,y) and (y,x) in median problems. Why?)
4. Suppose that town a has twice the population of town b which has twice the population of town c which has the same population as town d. Repeat Exercise 1 using the weights described.
5. Simplify the method of Section 2 for finding an absolute center for the special case when the graph under consideration is a tree.
6. Show that the interior of a directed arc can never contain an absolute center or an absolute median.
7. How much can the length of edge (1,4) in Fig. 8.4 increase without changing the location of the center of this graph? Median? How much can the length of edge (2,3) increase without changing the location of the absolute canter of this graph? Absolute median?

REFERENCES

Christofides, N. and P. Viola, 1971. The Optimum Location of Multi-centres of a Graph, *O. R. Quart.*, 22, pp. 45-54.

Goldman, A. J., 1969. Optimum Locations for Centers in a Network, *Transp. Sci.*, 4, pp. 352-360.

Hakimi, S. L., 1964. Optimum Locations of Switching Centers and the Absolute Centers and Medians of a Graph, *ORSA*, 12, pp. 450-459.

Hakimi, S. L., 1965. Optimum Distribution of Switching Centers in a Communications Network and Some Graph Theoretic Problems, *ORSA*, 13, pp. 462-475.

Levy, J., 1967. An Extended Theorem for Location in a Network, *O.R. Quart.*. 18, pp. 433-442.

Marsten, R., 1972. An Algorithm for Finding Almost All of the Medians of a Network, DP #23, Center for Mathematical Studies in Economics and Management Science, Northwestern University, Evanston, November 1972

Minieka, E., 1970. The m-Center Problem, *SIAM Rev.*, 12, pp. 138-139.

Minieka, E., 1977. The General Centers and Medians of a Graph, *ORSA*, 25, pp. 641-650.

ReVelle, G., and R. Swain, 1970. Central Facility Location, *Geographical Analysis*, 2, no. 1, 30-42.

Toregas, C., C. ReVelle, R. Swain, and L. Bergman, 1971. The Location of Emergency Facilities, *ORSA*, 19, pp. 1363-1373.

Wendell, R. E., and A. P. Hurter, Jr., 1973. Optimal Locations on a Network, *Transp. Sci.*, 7, pp. 18-33.

Chapter 9

PROJECT NETWORKS

9.1 CRITICAL PATH METHOD (CPM)

Large projects such as the construction of a building, development of an accounting information system, graduating from college in four years, or even preparation of a dinner party, involve a large number of different activities. Some of these activities can be performed at the same time; others can be performed only after certain other activities have been completed. For example, the landscaping and interior plastering of a building under construction can be performed simultaneously; however, the walls cannot be erected until after the foundation has been laid. In a four-year college program, each course may be regarded as an activity. Some courses may be taken concurrently with others while other courses serve as prerequisites to more advanced courses. When preparing a dinner, you can set the table while the roast is in the oven. However, you cannot cook the potatoes until after they have been washed.

Since we must contend with precedence relations between activities, as described above, and since each activity in a project requires a certain amount of time, the project manager is confronted with the dilemma of carrying out the activities in the best possible way so that the project finishes on time. For example, if a contractor delays too long in laying the building foundation, he will fail to meet the construction deadline. If a student delays taking basic courses, he will not graduate on time. If you spend all your time setting the table and forget to put the roast in the oven, dinner will be late. In other words, the project manager must determine which activities are *critical* to the on-time completion of the project.

A project may be represented by a graph, called a *project network*, as follows:

1. List each activity in the project
2. List the time required to perform each activity
3. List the immediate prerequisite activities of each activity
4. Represent each activity by an arc. Arrange the arcs into network so that only the activities that immediately precede the activity represented by arc (x,y) are incident into vertex x. If necessary, create dummy arcs that represent no activity to achieve this.

EXAMPLE 1. Suppose that the construction of a new home requires the following activities:

Activity	Time	Prerequisites	Arc
1. Clear land	1	None	(a,b)
2. Lay foundation	4	Clear land	(b,c)
3. Erect walls	4	Lay foundation	(c,d)
4. Wire	3	Erect walls	$(d,e)_1$
5. Plaster	4	Wire roof	(e,f)
6. Landscape	6	Lay foundation	(d,g)
7. Interior work	4	Plaster roof	(f,g)
8. Roof	5	Erect walls	$(d,e)_2$

Observe that the interior work activity requires as prerequisite activities both plastering and roofing. However, roofing is also a prerequisite activity for plastering. Hence, it is redundant to list roofing as a prerequisite activity for interior work. To ease the analysis of this project, roofing should be removed as a prerequisite activity for interior work.

9.1 Cricital Path Method (CPM)

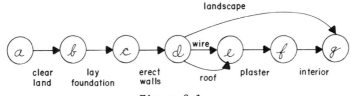

Figure 9.1
Project Network for Building a House

The project network is shown in Fig. 9.1. Observe that there are two parallel activities joining vertices d and e. In practice, parallel arcs are discouraged and often replaced by either a composite arc representing a compositive activity (in this case, wiring and roofing) or by replacing one of the parallel arcs by two arcs in series. To ease the ensuing presentations, we shall not follow these conventions.

EXAMPLE 2. Consider the following abstract project:

Activity	Prerequisites
A	None
B	A
C	A
D	A
E	B,C
F	B,C,D
G	E,F

Observe that the prerequisites of activity E are a proper subset of the prerequisites of activity F. In this situation, we need a dummy activity to depict these precedence relationships (see Fig. 9.2). Arc (c,d) is a dummy arc; it is needed to insure that activities B and C precede activity F. Dummy activities are always assigned zero time.

EXAMPLE 3. A student must complete courses in (a) calculus (2 terms), (b) statistics (3 terms), (c) linear programming (1 term), (d) non-linear programming (1 term), and (e) stochastic programming (1 term),

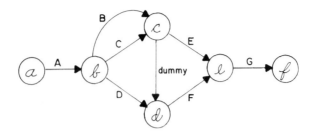

Figure 9.2
Project Network with Dummy Arc

before he can graduate in operations research. Needless to say, he cannot enroll in Calculus II until he has successfully completed Calculus I; he cannot enroll in Statistics III until he has completed Statistics II and Calculus II, and he cannot enroll in Statistics II until he has successfully completed Statistics I which requires Calculus I. There are no prerequisites for linear programming; however, the prerequisites for nonlinear programming are Calculus II and linear programming. The prerequisites for stochastic programming are Calculus II, Statistics III, and linear programming.

The project network for this study program is shown in Fig. 9.3. Note that a dummy arc (c,d) is required since Statistics II requires both Statistics I and Calculus II, but nonlinear programming requires Calculus II and linear programming.

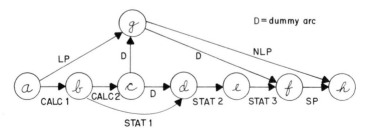

Figure 9.3
Project Network for Operations Research
Course Program

9.1 Critical Path Method (CPM)

The student can complete this program in as little as 5 terms if he passes Calculus I, Statistics I, Statistics II, Statistics III and Stochastic Programming in 5 successive terms. Any delay in this sequence will delay the completion of his program by an equal amount of time.

Any graph depicting the relationships between the activities of a project is called a *project network*. The arcs of a project network are always directed and represent either real or dummy activities. The vertices of a project network are called *events*. An event is said to have been *completed* when all the activities directed into it have been completely performed.

A project network cannot contain a circuit. Otherwise, if there were a circuit, say (a,b), (b,c), ..., (r,s), (s,a), then event a could not be completed until event s had been completed, and event s could not be completed until event r had been completed, etc., which implies that the project can never be completed. Any real project must be capable of being completed, and hence, its project network cannot contain any circuits.

Consider any project network $G = (X,A)$ with event set X and activity set A. To simplify all future developments, assume that network G has exactly one event with no activities directed into it and exactly one event with no activities directed out of it. These two events can be regarded respectfully as the *start* and *finish* events analogous to the source and sink in a flow flow.

Since there are no circuits in G, the events can be numbered 1, 2, ..., so that for any activity (i,j), we have $i < j$. This is achieved as follows.

Event Numbering Algorithm

Step 1: Give the start event number 1.

Step 2: Give the next number to any unnumbered event whose predecessor events are each already numbered. (Such an event exists since there are no circuits in the network.) Repeat Step 2 until all events have been numbered. [Note that the finish event will always receive the last (highest) number.]

EXAMPLE 2 (Event Numbering Algorithm). Number the events in the project network in Fig. 9.2. Event a is the start event and according to Step 1 of the algorithm is given number 1. Proceeding to Step 2, event b has only numbered preceding events (i.e., event a); hence, event b is given number 2. Next, event c is given number 3; event d is given number 4; event e is given number 5, and finally finish event f is given number 6. Observe that in this example there was no choice as to which event would receive the next number.

The events in the project network in Fig. 9.4 can be numbered in a variety of different ways:

Event	First numbering	Second numbering	Third numbering
a	1	1	1
b	3	2	2
c	4	4	3
d	2	3	4
e	5	5	5

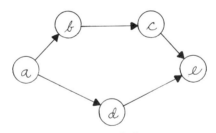

Figure 9.4
Project Network with Three Possible
Event Numberings

Denote the amount of time requires to perform activity (x,y) by $t(x,y) \geq 0$. If activity (x,y) is a dummy activity, then let $t(x,y) = 0$.

Next, let us analyze a project network to determine how early the project can be completed and which activities are critical to the on-time completion of the project.

9.1 Critical Path Method (CPM)

For each event $x \in X$, let $E(x)$ denote the *earliest time* that event x can possibly be completed. Let $L(x)$ denote the *latest time* at which event x can be completed such that the project will still be completed on time.

For example, in Fig. 9.3, $E(b) = 1$ since event b can be completed as easly as the end of the first term. If the completion deadline for this project is the end of the fifth term, $L(b) = 1$ since the latest that event b can be completed is the end of the first term; otherwise, Statistics I, II and III and stochastic programming could not be completed by the end of the fifth term.

As another example, suppose that event 14 has only three immediate predecessor events, namely events 5, 8, and 9, where $E(5) = 4$, $E(8) = 7$ and $E(9) = 6$ (see Fig. 9.5). Event 14 cannot be completed earlier than time 10 since $E(5) + t(5,14) = 4 + 6 = 10$. Also, event 14 cannot be completed earlier than time 11 since $E(8) + t(8,14) = 7 + 4 = 11$. Moreover, event 14 cannot be completed earlier than time 9 since $E(9) + t(9,14) = 6 + 3 = 9$. Thus, $E(14) = \max\{10,11,9\} = 11$.

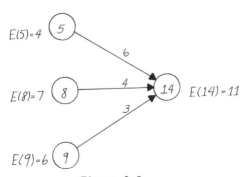

Figure 9.5
Calculating Earliest Event Times

In general, we see that

$$E(j) = \max_{i:(i,j) \in A} \{E(i) + t(,i,j)\} \tag{1}$$

With equation (1) as motivation, we may now state the *earliest event time algorithm*.

Earliest Event Time Algorithm

Step 1: Number the events 1, 2, ..., n = $|X|$ so that all activities (i,j) have i < j. Use the event numbering algorithm to accomplish this. Let $E(1) = 0$.

Step 2: For j = 2, 3, ..., n, let

$$E(j) = \max_{i:(i,j) \in A} \{E(i) + t(i,j)\}$$

EXAMPLE 5 (Earliest Event Time Algorithm). Calculate the earliest event times for the project network shown in Fig. 9.6. The events are already numbered 1 through 7.

Step 1: $E(1) = 0$

Step 2: $E(2) = E(1) + t(1,2) = 0 + 4 = 4$
$E(3) = \max[E(1) + t(1,3), E(2) + t(2,3)]$
$= \max[4 + 1, 0 + 3] = 5$
$E(4) = E(1) + t(1,4) = 0 + 4 = 4$
$E(5) = \max[E(2) + t(2,5), E(3) + t(3,5)]$
$= \max[4 + 7, 5 + 4] = 11$
$E(6) = \max[E(4) + t(4,6), E(5) + t(5,6)]$
$= \max[4 + 2, 11 + 1] = 12$
$E(7) = \max[E(2) + t(2,7), E(5) + t(5,7), E(6) + t(6,7)]$
$= \max[4 + 8, 11 + 3, 12 + 4] = 16$

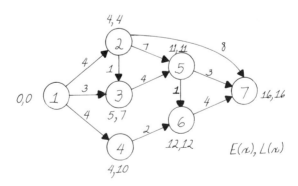

Figure 9.6
Example of Event Time Algorithms

9.1 Critical Path Method (CPM)

Consequently, the project cannot be completed any earlier than time 16.

Next, let us calculate the latest event times. For example, consider event 17 in Fig. 9.7. Event 17 is a predecessor to exactly three events, namely events 20, 24 and 29. Event 17 cannot be completed later than time 11; otherwise, event 20 would be delayed past its latest time which is 16. Similarly, event 17 cannot be completed later than time 15 since L(24) - t(17,24) = 19 - 4 = 15. Also, event 17 cannot be completed later than time 14 since L(29) - t(17,29) = 22 - 8 = 14.

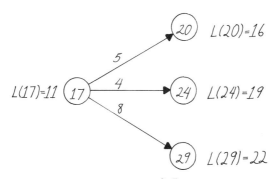

Figure 9.7
Calculating Latest Event Times

In general, we see that

$$L(i) = \min_{j:(i,j) \in A} \{L(j) - t(i,j)\} \qquad (2)$$

With equation (2) as motivation, we may now state the *latest event time algorithm*.

Latest Event Time Algorithm

Step 1: Number the events 1, 2, ..., n = |X| so that all activities (i,j) have i < j. Use the event numbering algorithm to accomplish this.

Let L(n) equal the time at which the project must be completed. [In any realistic situation L(n) ≥ E(n).]

Step 2: For i = n - 1, n - 2, ..., 1, let

$$L(i) = \min_{j:(i,j) \in A} \{L(j) - t(i,j)\}$$

EXAMPLE 6 (Latest Event Time Algorithm). Compute the latest event times for the project network shown in Fig. 9.6. Let $L(7) = E(7) = 16$ which indicates that the project must be finished as early as possible, i.e., time 16.

$L(7) = 16$
$L(6) = L(7) - t(6,7) = 16 - 4 = 12$
$L(5) = \min\{L(7) - t(5,7), L(6) - t(6,5)\}$
$\quad = \min\{16 - 3, 12 - 1\} = 11$
$L(4) = L(6) - t(4,6) = 12 - 2 = 10$
$L(3) = L(5) - t(3,5) = 11 - 4 = 7$
$L(2) = \min\{L(7) - t(2,7), L(5) - t(2,5), L(3) - t(2,3)\}$
$\quad = \min\{16 - 8, 11 - 7, 7 - 1\} = 4$
$L(1) = \min\{L(4) - t(1,4), L(3) - t(1,3), L(2) - t(1,2)\}$
$\quad = \min\{10-4, 7-3, 4-4\} = 0$

Consequently, the project must begin at time 0 in order to finish at time 16.

It follows directly from the latest event time algorithm that if the latest completion time of the entire project $L(n)$ is increased by t time units, then the latest time of every event will also be increased by t units.

The earliest time $E(x)$ of event x can be interpreted as the length of the longest path from the start event to event x. Similarly, $L(n) - L(x)$ can be interpreted as the length of the longest path from event x to the finish event. Lastly, observe that if $L(n) \geq E(n)$, then $L(x) \geq E(x)$ for all events x.

The earliest event time algorithm requires only one addition for each activity in the project and one maximization for each event in the project, except event 1. Similarly, the latest event time algorithm requires only one subtraction for each activity in the project and one minimization for each event in the project, except for event n.

9.1 Critical Path Method (CPM)

Consider any activity (x,y). What is the maximum amount of time that can be allotted to activity (x,y) without delaying the on-time completion of the entire project? Activity (x,y) may start as early as time $E(x)$ and may finish as late as time $L(y)$. Hence, at most $L(y) - E(x)$ time periods may be allotted to the performance of activity (x,y) without delaying the on-time completion of the entire project. Consequently, the maximum delay that can be tolerated in activity (x,y) is $L(y) - E(x) - t(x,y) \geq 0$. The quantity

$$L(y) - E(x) - t(x,y) \qquad (3)$$

is called the *total float* of activity (x,y). Obviously, if the total float of an activity equals zero, then any delay in the performance of this activity will delay the on-time completion of the entire project by an equal amount.

How much time can be allotted to the performance of activity (x,y) without imposing any additional time constraints on the activities that are performed after (x,y)? In this case, activity (x,y) must be completed by time $E(y)$. Since activity (x,y) may begin as early as time $E(x)$, it follows that at most $E(y) - E(x)$ time periods may be allotted to the performance of activity (x,y) without imposing any additional time constraints on the activities that follow (x,y). The quantity

$$E(y) - E(x) - t(x,y) \qquad (4)$$

is called the *free float* of activity (x,y). The free float of activity (x,y) equals the maximum delay that can occur in the performance of activity (x,y) without effecting any activity that follows (x,y). From equation (1), it follows that the free float is always nonnegative.

How much time can be allotted to the performance of activity (x,y) without imposing any additional time constraint on any other activity in the project? In order not to impose any additional requirements on any other activity in the project, activity (x,y) must begin as late as possible and be completed as easly as possible. Thus, activity (x,y) would have to begin at time $L(x)$ and end at time $E(y)$. Thus, at most $E(y) - L(x)$ time periods can be allotted to the performance of activity (x,y). The quantity

$$E(y) - L(x) - t(x,y) \tag{5}$$

is called the *independent float* of activity (x,y). The independent float of activity (x,y) can be interpreted as the maximum delay that can occur in the performance of activity (x,y) without imposing any additional time restriction on any other activity in the project. A negative value for an independent float indicates that any delay will effect the flexibility of other activities in the project.

How are the three kinds of floats related? Since $L(x) \geq E(x)$ for all events x, it follows from statements (3)-(5) that for each activity (x,y)

$$\text{Total float} \geq \text{Free float} \geq \text{Independent float} \tag{6}$$

The following table gives the three floats for each activity in the project network in Fig. 9.6.

Activity	Total Float	Free Float	Independent Float	
(1,2)	4 - 0 - 4 = 0	4 - 0 - 4 = 0	4 - 0 - 4 = 0	(critical)
(1,3)	7 - 0 - 3 = 4	5 - 0 - 3 = 2	5 - 0 - 3 = 2	
(1,4)	10 - 0 - 4 = 6	4 - 0 - 4 = 0	4 - 0 - 4 = 0	
(2,3)	7 - 4 - 1 = 2	5 - 4 - 1 = 0	5 - 4 - 1 = 0	
(2,5)	11 - 4 - 7 = 0	11 - 4 - 7 = 0	11 - 4 - 7 = 0	(critical)
(2,7)	16 - 4 - 8 = 4	16 - 4 - 8 = 4	16 - 4 - 8 = 4	
(3,5)	11 - 5 - 4 = 2	11 - 5 - 4 = 2	11 - 7 - 4 = 0	
(4,6)	12 - 4 - 2 = 6	12 - 4 - 2 = 6	12 - 10 - 2 = 0	
(5,6)	12 - 11 - 1 = 0	12 - 11 - 1 = 0	12 - 11 - 1 = 0	(critical)
(5,7)	16 - 11 - 3 = 2	16 - 11 - 3 = 2	16 - 11 - 3 = 2	
(6,7)	16 - 12 - 4 = 0	16 - 12 - 4 = 0	16 - 12 - 4 = 0	(critical)

An activity is called *critical* if any delay in the performance of the activity delays the on-time completion of the entire project. In other words, a critical activity is any activity whose total float equals zero.

9.1 Critical Path Method (CPM)

It is vital for the project manager to identify all critical activities so that he can guard against any delays in these activities as these delays would delay the on-time completion of the entire project. Delays less than the total float may occur in noncritical activities without delaying the on-time completion of the entire project.

Recall that $E(n)$ equals the length of a longest path from the start event to the finish event of the project. If $E(n) = L(n)$, then every activity on a longest path from the start event to the finish event is critical. A path consisting entirely of critical activities is called a *critical path*.

For example in the project network shown in Fig. 9.6, activities (1,2), (2,5), (5,6), and (6,7) each have total float equal to zero. Thus, each of these activities is a critical activity. Notice that these activities form a path from 1 to 7 whose total length is $4 + 7 + 1 + 4 = 16$. This is the longest path in the project network. Any delays along the activities of this path will result in an equal delay in the on-time completion of the entire project.

As a further example, consider the project network shown in Fig. 9.8. Notice that path (1,2), (2,5) and path (1,3), (3,5) each have total length equal to 8 time units. Thus, both of these paths are critical.

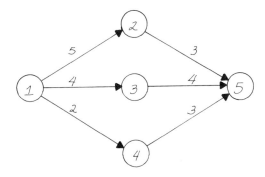

Figure 9.8
Project with Multiple Critical Paths

Activities (1,4) and (4,5) are not critical as limited delays in both of these activities can be tolerated without delaying the completion of the entire project by time 8. How much can activity (1,4) be delayed without delaying the completion of the project beyond time period 8?

Up to now, we have assumed that $t(x,y)$ is known with certainty for all activities (x,y). Obviously, this is hardly realistic. Can we ever be certain how much time a machine shop job will require? Can we be certain that we shall complete a course in one term? Can we be certain how long the landscaping will require? Can we be certain how long a 5 lb. roast will require?

To deal with the problem of uncertainty in activity times, a technique known as *Program Evaluation Review Technique (PERT)* was developed. In essence, PERT is the same as the critical path method (CPM) described except activity times are replaced with expected activity times in PERT. To calculate the expected time of an activity, PERT requires that three time estimates be made. They are

A, the *optimistic activity time*

B, the *realistic activity time*

C, the *pessimistic activity time*

The expected activity time is estimated to be

$$\frac{A}{6} + \frac{4}{6}B + \frac{C}{6} \tag{7}$$

In other words, a weighted average of the three time estimates is taken, where A and C have 1/6 weights and B has weight 4/6.

The variance of each activity time is taken as

$$\left(\frac{C - A}{6}\right)^2 \tag{8}$$

The critical path method (CPM) can now be applied to the project network with activity times replaced by expected activity times as computed in expression (7). Now $E(x)$ denotes the *expected* earliest time of event x, and $L(x)$ denotes the *expected* latest time of event x. Hence, $E(n)$ denotes the expected earliest completion time of the entire project.

9.1 Critical Path Method (CPM)

The *actual* earliest completion time of the entire project is assumed to be a normally distributed random variable whose mean equals $E(n)$ and whose variance equals the sum of the variances of the activities in the longest path from event 1 to event n in the project network. (If there is more than one such path as in Fig. 9.8, then the largest variance of any such path is used.) This assumption about the mean, variance and normal distribution of the actual completion time permits us to make probability statements about the actual completion time. See any introductory statistics text for the details of how to make probability statements about a normally distributed random variable.

The PERT user should bear in mind that this normal assumption is weak if the activity times tend not to be statistically independent of one another. Moreover, the theoretical justifications for expressions (7) and (8) rest on some tenuous connections between activity times and the beta distribution. Nonetheless, PERT has received widespread industrial acceptance.

9.2 MINIMUM COST ACTIVITY TIMES

If a project must be completed at the earliest possible time, then the amount of time that can be allotted to the performance of each activity in the project is quite inflexible. In fact, critical activities, as we saw in Sec. 9.1, cannot tolerate any delays. A noncritical activity (x,y) may be delayed only so long as it finishes no later than time $L(y)$. Consequently, a project manager is faced with the decision of how to schedule optimally the delays among the noncritical activities. This section is devoted to a model developed by Fulkerson (Ford and Fulkerson, 1962) for solving such a problem at minimum cost.

Suppose that the time $t(x,y)$ allotted to the performance of activity (x,y) must satisfy the following upper and lower bound requirements:

$$0 \leq r(x,y) \leq t(x,y) \leq s(x,y) \tag{9}$$

Also, suppose the total cost incurred for activity (x,y) is

$$K(x,y) - k(x,y)t(x,y) \tag{10}$$

where $K(x,y)$ is any constant, and $k(x,y)$ is a positive constant. In other words, it costs $k(x,y)$ to decrease the performance time by one unit. For many situations, a linear cost function (10) is fairly realistic. For example, activity time might be decreased by adding additional workers paid at the same rate.

How can we assign a performance time $t(x,y)$ to each activity (x,y) so that the project will be completed by time T (i.e., on time) with the smallest possible total cost?

A solution to this minimum cost activity time problem consists of selecting an optimum completion time $p(x)$ for each event x such that

$$p(1) = 0 \quad p(n) = T \tag{11}$$
$$p(y) - p(x) \geq r(x,y) \quad \text{[for all } (x,y)\text{]} \tag{12}$$

Then, the performance time for each activity (x,y) is made as large as possible, i.e., $t(x,y)$ is set equal to

$$\min\{p(y) - p(x), s(x,y)\} \tag{13}$$

Hence, to solve the minimum cost activity time problem, we need only locate optimum event numbers $p(x)$ that satisfy conditions (11) and (12).

The minimum cost activity time problem is represented by the following linear programming problem:

Minimize
$$\sum_{(x,y)} [K(x,y) - k(x,y)t(x,y)] \tag{14}$$

such that
$$p(x) + t(x,y) \leq p(y) \quad \text{[for all } (x,y)\text{]} \tag{15}$$
$$r(x,y) \leq t(x,y) \quad \text{[for all } (x,y)\text{]} \tag{16}$$
$$t(x,y) \leq s(x,y) \quad \text{[for all } (x,y)\text{]} \tag{17}$$
$$p(n) - p(1) \leq T \tag{18}$$

From the project network $G = (X,A)$, construct the graph $G' = (X,A')$ as follows. Replace each activity $(x,y) \in A$ by two arcs

9.2 Minimum Cost Activity Times

$(x,y)_1 \in A'$ and $(x,y)_2 \in A'$. Also create a "return" arc $(n,1) \in A'$. Let $a(x,y)_i$ denote the cost of sending one flow unit across arc $(x,y)_i$, $i = 1, 2$, where

$$a(x,y)_1 = -s(x,y)$$
$$a(x,y)_2 = -r(x,y) \qquad (19)$$

and

$$a(n,1) = T$$

Let $c(x,y)_i$ denote the capacity on arc $(x,y)_i$, $i = 1, 2$, where

$$c(x,y)_1 = k(x,y)$$
$$c(x,y)_2 = \infty \qquad (20)$$
$$c(n,1) = \infty$$

THEOREM 9.1 Apply the out-of-kilter algorithm of Sec. 4.4, to find a minimum cost flow in graph G'. Let $p_t(x)$ denote the value of the dual variable for vertex x at the termination of the out-of-kilter algorithm.

Then, $p(x) = -(p_t(x) - p_t(1))$ for all $x \in X$ are optimal values for the event completion times.

Proof: Since $\sum_{(x,y)} K(x,y)$ is a constant, expression (14) can be replaced by

Minimize

$$\sum_{(x,y)} k(x,y) t(x,y) \qquad (21)$$

Observe that expressions (15)-(18), (21) form a linear programming problem whose decision variables are $t(x,y)$, $(x,y) \in A$, and $p(x)$, $x \in X$. Write the dual to this linear programming problem. Let $f(x,y)$ be the dual variable corresponding to constraint (15) for activity (x,y). Let $h(x,y)$ be the dual variable corresponding to constraint (16) for activity (x,y). Let $g(x,y)$ be the dual variable corresponding to constraint (17) for activity (x,y). Let v denote the dual variable for constraint (18).

The dual linear programming problem is

Minimize

$$Tv - \sum_{(x,y)} r(x,y)h(x,y) + \sum_{(x,y)} s(x,y)g(x,y) \qquad (22)$$

such that

$$\sum_y [f(x,y) - f(y,x)] = \begin{cases} 0 & \text{For } x \neq 1, n \\ v & \text{For } x = 1 \\ -v & \text{For } x = n \end{cases} \qquad (23)$$

$$f(x,y) + g(x,y) - h(x,y) = k(x,y) \quad [\text{for all } (x,y)] \qquad (24)$$

$$f(x,y) \geq 0 \quad [\text{for all } (x,y)] \qquad (25)$$

$$g(x,y) \geq 0 \quad [\text{for all } (x,y)] \qquad (26)$$

$$h(x,y) \geq 0 \quad [\text{for all } (x,y)] \qquad (27)$$

Since $g(x,y) - h(x,y) = k(x,y) - f(x,y)$, and since $r(x,y) \leq s(x,y)$, it follows that in any optimal solution to the dual linear programming problem

$$g(x,y) = \max\{0, k(x,y) - f(x,y)\} \qquad (28)$$

$$h(x,y) = \max\{0, f(x,y) - k(x,y)\} \qquad (29)$$

Using equations (28)-(29), the dual objective function (22) becomes

Minimize

$$Tv + \sum[s(x,y) \max\{0, k(x,y) - f(x,y)\}] \\ - \sum[r(x,y) \max\{0, f(x,y) - k(x,y)\}] \qquad (30)$$

Observe that as $f(x,y)$ increases from 0 to $k(x,y)$, objective function (30) decreases at a rate of $s(x,y)$. As $f(x,y)$ increases above $k(x,y)$, objective function (30) decreases at a rate of $r(x,y)$. Since objective function (30) is to be minimized, increases of $f(x,y)$ from 0 to $k(x,y)$ have a more beneficial effect on the objective function than do increases of $f(x,y)$ beyond $k(x,y)$, since $r(x,y) \leq s(x,y)$.

Hence, we may regard activity (x,y) as an arc with the property that the first $k(x,y)$ flow units using this arc cost $s(x,y)$ apiece and the remaining flow units to use this arc cost $r(x,y)$ apiece.

9.2 Minimum Cost Activity Times

Consequently, replace each activity (x,y) by two arcs $(x,y)_1$ and $(x,y)_2$. Let

$$f(x,y) = f(x,y)_1 + f(x,y)_2 \tag{31}$$

Define the costs and capacity of each arc $(x,y)_1$ and $(x,y)_2$ as in equations (19) and (20).

The dual linear programming problem can now be rewritten as

Minimize

$$Tv - \sum_{(x,y)_i} a(x,y)_i f(x,y)_i \tag{32}$$

such that

$$\sum_y \sum_{i=1,2} [f(x,y)_i - f(y,x)_i] = \begin{cases} 0 & \text{For } x \neq 1, n \\ v & \text{For } x = 1 \\ -v & \text{For } x = n \end{cases} \tag{33}$$

$$0 \leq f(x,y)_i \leq c(x,y)_i \quad [\text{for all } (x,y)_i] \tag{34}$$

The dual linear programming problem (32)-(34) can be transformed into an even more recognizable form as follows: Add a "return" arc from n to 1. Let the cost and capacity of arc (n,1) be defined in equations (19)-(20). Expressions (31)-(34) become

Minimize

$$\sum_{(x,y)} a(x,y) f(x,y) \tag{35}$$

such that

$$\sum_y [f(x,y) - f(y,x)] = 0 \quad (\text{for all } x) \tag{36}$$

$$0 \leq f(x,y) \leq c(x,y) \quad [\text{for all } (x,y)] \tag{37}$$

where the arc subscripts have been suppressed for brevity. Linear programming problem (35)-(37) is none other than the minimum cost flow problem found in Chap. 4, expressions (24)-(26) for graph G'.

Every flow unit must travel from vertex 1 to vertex n and return to vertex 1 via the return arc (n,1). A flow unit incurs a cost +T on the return arc and incurs only negative costs on the arcs it travels from 1 to n [see equation (19)]. Since the objective function (35) is to be minimized, it follows that a flow unit is dispatched to make the round trip only if it incurs a total cost not exceeding -T on its travels from vertex 1 to vertex n.

Problem (35)-(37) can be solved by applying the out-of-kilter algorithm to graph G'. See Sec. 4.4. Upon termination, the out-of-kilter algorithm will produce a feasible flow for graph G' that satisfies conditions (36)-(37) above and will also produce dual vertex values $p_t(x)$, $x \in X$ that satisfy the complementary slackness conditions

$$p_t(y) - p_t(x) < a(x,y) \Rightarrow f(x,y) = 0 \quad (38)$$

$$p_t(y) - p_t(x) > a(x,y) \Rightarrow f(x,y) = c(x,y) \quad (39)$$

for all $(x,y) \in A'$. [See Chap. 4, conditions (33)-(34), where $l(x,y) = 0$ for all (x,y).]

If each activity (x,y) were assigned its maximum possible time $s(x,y)$, then the total time required by the project would exceed T. [Otherwise, the original problem is trivial since an optimum solution is $t(x,y) = s(x,y)$ for all (x,y).] Since $a(x,y)_1 = -s(x,y)$ for all (x,y) it follows that at least some flow units travel from vertex 1 to vertex n and return along arc $(n,1)$ in every optimum solution to problem (35)-(37). Thus, $f(n,1) > 0$, and from (38)-(39) it follows that

$$p_t(1) - p_t(n) = a(n,1) = T \quad (40)$$

Without loss of generality, we may assume that $p_t(1) = 0$. If this is not the case, subtract $p_t(1)$ from every $p_t(x)$. This will not alter the validity of conditions (38)-(39). Hence, $p_t(n) = -T$.

Since equation (36) is the negative of equation (25) in Chap. 4, the dual vertex variable values produced by the out-of-kilter algorithm are the negative of the dual vertex variables required by problem (35)-(37). Hence, the optimum dual vertex values produced by the out-of-kilter algorithm must be multiplied by -1.

9.2 Minimum Cost Activity Times

Clearly, any optimum flow for problem (35)-(37) corresponds to an optimum flow for problem (22)-(27) where $f(x,y) = f(x,y)_1 + f(x,y)_2$. To complete the proof, we need only show that the values $p(x) = -[p_t(x) - p_t(1)]$ are feasible for the original problem (15)-(18) and (21). Secondly, we must show that the values $p(x)$ satisfy the complementary slackness conditions existing between the original problem (15)-(18) and (21) and its dual (22)-(27). These complementary slackness conditions are

$$p(y) - p(x) - t(x,y) > 0 \Rightarrow f(x,y) = 0 \qquad (41)$$
$$r(x,y) < t(x,y) \Rightarrow h(x,y) = 0 \qquad (42)$$
$$t(x,y) < s(x,y) \Rightarrow g(x,y) = 0 \qquad (43)$$

To show feasibility, note that arc $(x,y)_2$ can never have a capacity flow since its capacity is ∞. Hence,

$$p_t(y) - p_t(x) \leq a(x,y)_2 = -r(x,y)$$

Thus,

$$p(y) - p(x) \geq r(x,y)$$

Consequently, the values $p(x)$ provide a feasible solution for the original problem where

$$t(x,y) = \min\{p(y) - p(x), s(x,y)\}$$

Next, let us verify complementary slackness condition (41). Note from expression (13) that

$$p(y) - p(x) - t(x,y) > 0 \Rightarrow t(x,y) = s(x,y)$$

Thus,

$$-p_t(y) + p_t(x) - a(x,y)_1 > 0$$

and by (38), it follows that $f(x,y)_1 = 0$. Thus, $f(x,y)_2 = 0$, and $f(x,y) = 0$, and condition (41) is satisfied.

To verify complementary slackness condition (42), note that

$$r(x,y) < t(x,y) \Rightarrow r(x,y) < p(y) - p(x)$$
$$\Rightarrow p_t(y) + p_t(x) > r(x,y) = -a(x,y)_2$$

Thus, from (38), it follows that $f(x,y)_2 = 0$. Hence,

$$f(x,y) = f(x,y)_1 \le k(x,y)$$

and from conditions (27) and (29), it follows that $h(x,y) = 0$. Thus, condition (42) is verified.

Complementary slackness condition (43) follows in a similar way. Q.E.D.

EXAMPLE. Use Theorem 1 to calculate the optimum times to assign to each activity in the project network G shown in Fig. 9.9. The corresponding graph G' is shown in Fig. 9.10. Observe that G' is merely G with all its arcs repeated and, in addition, a return arc.

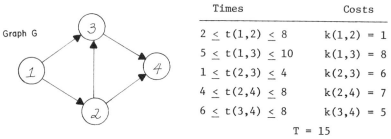

	Times	Costs
	$2 \le t(1,2) \le 8$	$k(1,2) = 1$
	$5 \le t(1,3) \le 10$	$k(1,3) = 8$
	$1 \le t(2,3) \le 4$	$k(2,3) = 6$
	$4 \le t(2,4) \le 8$	$k(2,4) = 7$
	$6 \le t(3,4) \le 8$	$k(3,4) = 5$
		$T = 15$

Figure 9.9
Project Network G with Variable Activity Times

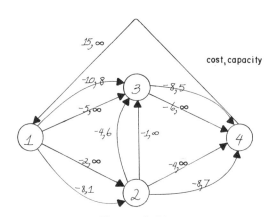

Figure 9.10
Graph G'

9.2 Minimum Cost Activity Times

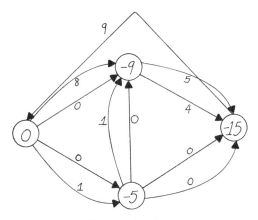

Figure 9.11

Optimum Results of the Out-of-Kilter Algorithm
(Flow values and vertex numbers are given.)

The results of the out-of-kilter algorithm are given in Fig. 9.11. It is left to the reader that the vertex numbers p(x) and the flow resulting from the out-of-kilter algorithm satisfy the complementary slackness conditions (38)-(39).

The optimum event completion times are

$p(1) = -(p_t(1) - p_t(1)) = 0$

$p(2) = -(p_t(2) - p_t(1)) = -(-5 - 0) = 5$

$p(3) = -(p_t(3) - p_t(1)) = -(-9 - 0) = 9$,

$p(4) = -(p_t(4) - p_t(1)) = -(-15 - 0) = 15$

The optimum activity times are

$t(1,2) = \min\{s(1,2), p(2) - p(1)\} = \min\{8, 5 - 0\} = 5$

$t(1,3) = \min\{s(1,3), p(3) - p(1)\} = \min\{10, 9 - 0\} = 9$

$t(2,3) = \min\{s(2,3), p(3) - p(2)\} = \min\{4, 9 - 5\} = 4$

$t(2,4) = \min\{s(2,4), p(4) - p(2)\} = \min\{8, 15 - 5\} = 8$

$t(3,4) = \min\{s(3,4), p(4) - p(3)\} = \min\{8, 15 - 9\} = 6$

Observe that if a cost or capacity changes (i.e., if a time or cost in the original problem changes), the out-of-kilter algorithm can be initiated with the previous optimum solution as a starting solution. Often, for only small value changes, this initialization is more efficient than initializing the problem with a zero flow.

For example, the project manager may be instructed to finish the project two days earlier than previously planned. In this case, he must decrease T by two units. If the previous optimum solution from the out-of-kilter algorithm is used as the initial solution, then the return arc will be out of kilter by two units and all other arcs will be in kilter.

A detailed exposition of this method for finding optimum activity times can be found in Moder and Phillips (1970), Chap. 9.

9.3 GENERALIZED PROJECT NETWORKS

Up to now, we assumed that (a) all activities preceeding an event must be completed before any activities emanating from the event could be performed, and (b) all activities in the project must be performed.

For example, assumption (a) would be unnecessary when any one of several courses are the prerequisite for another course. Also, the arrival of any one of a number of checks would be sufficient for you to begin your shopping activity. Similarly, the success of any one of several grant proposals would suffice to finance a research project.

Assumption (b) would be unnecessary in a university program that allowed elective courses. Or, a milling job may have to undergo one, two, or three drillings depending upon the result of quality control tests. The results of a market survey may determine which type of advertising policy should be pursued.

Thus, we can see that many projects cannot be realistically described in terms of the confines of the project network defined in Sec. 9.1. For this reason, generalized project networks that avoid the above assumptions have been developed. A detailed description is available in Eisner (1962), Elmaghraby (1964), Pritsker and Happ (1966), Pritsker and Whitehouse (1966), and Pritsker (1977).

9.3 Generalized Project Networks

Unlike project networks which have only one kind of vertex called an event, generalized project networks have a variety of vertices, all commonly called *decision boxes*. A decision box or db is characterized by the conditions placed on the activities entering it and by the condition placed on the activities emanating from it.

Three different conditions can be placed on the activities entering a db

(a) "And input": All activities entering the db must be performed before the db is considered completed
(b) "Inclusive input": At least one activity entering the db must be performed before the db is considered completed
(c) "Exclusive input": Exactly one of the activities entering the db must be performed before the db is considered completed.

Two different conditions can be placed on the activities emanating from a db.

(a) "Deterministic output": All activities emanating from the db are to be performed once the db has been completed.
(b) "Probabilistic output": Exactly one of the activities emanating from the db is performed after the db has been completed.

Consequently, there are 3 × 2 = 6 different types of db's. Their pictorial representations are given in Fig. 9.12.

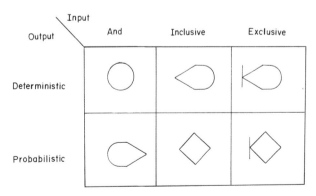

Figure 9.12
Six Types of Decision Boxes

In a project network, a time t(x,y) was specified for each activity (x,y). In a generalized activity network, both a time t(x,y) and a probability p(x,y) must be specifies for each activity (x,y). Probability p(x,y) denotes the chance that activity (x,y) will actually be performed once db x has been reached. If db x has a deterministic output, then p(x,y) must equal one and activity (x,y) is certainly performed. Moreover, the sum of the probabilities of the activities emanating from a probabilistic output db cannot exceed one.

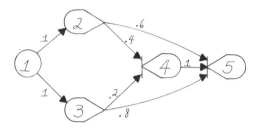

Figure 9.13
Generalized Project Network
(Activity numbers are probabilities.)

EXAMPLE 1. Consider the project whose generalized project network is shown in Fig. 9.13. Decision box 1 has a deterministic output; hence, both activities (1,2) and (1,3) will be performed. Thus, p(1,2) = 1 and p(1,3) = 1. Decision box 2 has an "and" input and is completed as soon as activity (1,2) has been performed. After db 2 has been completed, activity (2,5) will be performed with 60% probability, and activity (2,4) will be performed with 40% probability. Only one of the two activities (2,4) and (2,5) will occur. Decision box 4 is reached only if activity (2,4) or activity (3,4) is completed but not if both are completed. If neither (2,4) nor (3,4) are performed, then decision box 4 is never reached.

Moreover, it is possible that both activities (2,5) and (3,5) will be performed. In this case, db 5 is never reached.

9.3 Generalized Project Networks

EXAMPLE 2. A magazine welcomes contributions from would-be authors. Upon receipt of a manuscript, the magazine simultaneously submits it to two referees. A referee may reply in one of three different ways (reject, accept, or undecided) with respective probabilities 0.5, 0.4, and 0.1.

If the magazine receives at least one rejection recommendation, it rejects the manuscript. If the magazine receives acceptance recommendations from both referees, then it accepts the manuscript for publication. Otherwise, the magazine sends the manuscript to a third referee.

The generalized project network for this operation is shown in Fig. 9.14. The activities without names are dummy activities needed for the proper construction of all possible situations. Notice that all three possible db input types are present and that both possible db output types are present in the network.

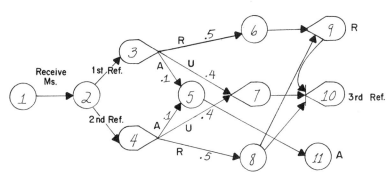

Figure 9.14

Generalized Project Network for Simultaneous Referees

(R = reject, A = accept, U = undecided)

Next, let us alter the situation. Suppose, instead, that the editor does not send the manuscript to the second referee unless the first referee returns an acceptance or undecided verdict. For this situation, the generalized project network is shown in Fig. 9.15.

In a project network, all events are eventually reached; it is merely a matter of time. The same is not necessarily true for a

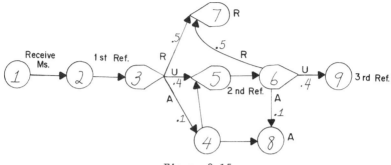

Figure 9.15
Generalized Project Network for Sequential Referees
(R = reject, A = accept, U = undecided)

generalized project network. Since not all activities need be performed, not all db's need be reached. For example, in Fig. 9.14, the project will eventually terminate at either db 9, 10, or 11.

Also, it can happen that the project will terminate, not with a db, but with an activity. For example, in Fig. 9.13, if both activities (2,4) and (3,4) are performed, then the project terminates without reaching db 4. However, this can occur only if exclusive input db's are present in the network.

Before proceeding into further analysis of generalized activity networks, let us note that it is often possible to simplify a network into an equivalent network with fewer activities. Such simplifications are possible for activities that occur in series or parallel combinations. These simplifications are shown in Fig. 9.16.

Because of the variety of possibilities for termination of a generalized project network, the project manager understandably wishes to know the probability that any db will in fact be reached and that any particular activity will in fact be performed. Moreover, it is important to know the expected time at which a db will be reached (supposing it is reached at all). (Note that in PERT, the actual time required to perform an activity was random, but we were certain that the activity would sooner or later be performed. In generalized activity networks the opposite case occurs: the time

9.3 Generalized Project Networks

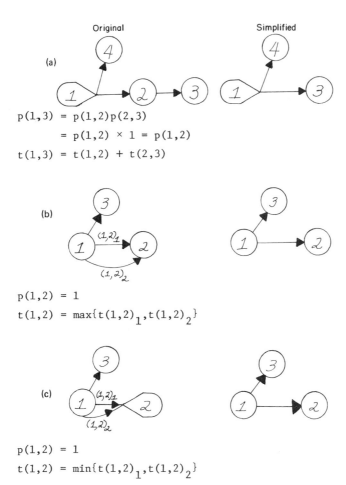

Figure 9.16
Network Simplifications

required to perform an activity is assumed to be nonrandom, but the activities that are performed are randomly selected.)

These probabilities and expected times are not easy to calculate. The difficulties in these calculations arise from the presence of probabilistic output db's. Only one of the set of activities emanating from a probabilistic output db may occur. Hence, the probabilities of the actually completing activities and db's following a probabilistic output db are not statistically independent, and consequently, we cannot use the convenient traditional probability rules which assume statistical independence.

For example, in Fig. 9.17, both activities (2,4) and (3,4) must be performed before db 4 is reached. There is a 0.6 × 0.5 = 0.3 probability that activity (2,4) is performed, There is a 0.4 × 0.4 = 0.16 probability that activity (3,4) is performed. Hence, we might hastily conclude that there is a 0.3 × 0.16 probability that db 4 is reached. However, upon closer inspection, we see that activities (2,4) and (3,4) are not statistically independent of one another. Both (2,4) and (3,4) cannot occur since, otherwise, both (1,2) and (1,3) must occur which is impossible. (Recall that only one activity emanating from a probabilistic output db may occur.) Thus, db 4 can never be reached.

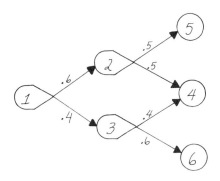

Figure 9.17
Statistical Dependence

Thus, the lack of statistical independence between activity probabilities leaves large networks virtually intractable. Even if the definition of the probabilistic output db were changed so that any number of activities could emanate from it (each with its own probability of occurrence), the same computational difficulties would arise. For example, several activities may all be descendents of one particular activity emanating from a probabilistic output db. Then, the occurrence of all these descendent activities would be

9.3 Generalized Project Networks   349

statistically dependent on one another, and we would encounter the
same computational difficulties as shown above.

As if the computational situation is not bad enough, many
projects generate generalized project networks that contain circuits.
For example, there might be an activity in the project that must be
repeated until it is performed correctly, such as a required course
in a university program. The presence of circuits complicates the
calculations even further. Some methods for simplifying circuits
into equivalent circuitless configurations can be found in Elmaghraby
(1964), Pritsker and Happ (1966), and Pritsker and Whitehouse (1966).

EXERCISES

1. Your company has decided to install a new, more efficient computer
   system. This will involve not only the phase in of the new
   system but also the phase out of the old system. All this must
   be accomplished so that at all times at least one system is
   operating. Moreover, canned programs must be converted to the
   new system's language and personnel must be trained to run the
   new system.
   Construct a detailed project network to represent this
   operation.

2. For the project network in Fig. 9.18
   (a) Calculate the earliest event times
   (b) Calculate the latest event times so that the project
       finished as early as possible
   (c) Find the critical path
   (d) Calculate the total float of each activity
   (e) Calculate the free float of each activity
   (f) Calculate the independent float of each activity

3. You discover that the time given for activity (3,5) in Fig.
   9.18 should have been 5. Update the results of Exercise 2 without repeating all calculations. How much can the performance
   time of activity (3,5) increase without altering the earliest
   completion time of this project?

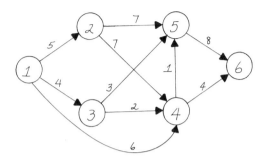

Figure 9.18
Exercise 9.2
(Numbers indicate activity times.)

4. In Fig. 9.19, the three numbers next to each activity are, respectively, the activity's optimistic time, realistic time, and pessimistic time. Use PERT to calculate

   (a) The expected time for each activity
   (b) The variance of each activity time
   (c) The earliest expected event times
   (d) The latest expected event times so that the project finishes at the earliest expected time
   (e) The critical path
   (f) The variance of the earliest completion time of the entire project

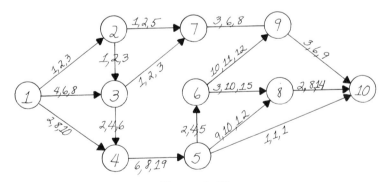

Figure 9.19
Exercise 9.4

Exercises 351

5. In Fig. 9.20, the numbers next to each activity indicate, respectively, the minimum time required to perform the activity, the maximum time required to perform the activity, and the cost in dollars incurred when the activity time is decreased by one time period.

    Calculate the optimum amount of time to be allotted to each activity so that the entire project is completed within 25 time periods.

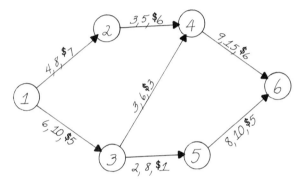

Figure 9.20
Exercise 9.5

6. Show that the earliest event time algorithm and the latest event time algorithm both fail when the network contain a circuit.

7. A drilling machine must drill two holes into an automobile part. If both holes are of the standard size, the part is considered finished. If the first hole drilled is below standard size, the second hole is drilled with extreme care on a special setting of the drilling machine. If the first hole is above standard size, then the driller must check if any of the pistons in his stock fit this hole. If so, the second hole is drilled at the usual setting. If not, there is a 60% chance that the quality control engineer will decide to discard the part; there is a 40% chance that the quality control engineer will decide to have the second hole drilled on the special setting.

    Construct a generalized project network for this operation.

8. A project consists of seven activities called A through G. The precedence relationships between these activities are

Activity	Predecessors
A	None
B	A
C	None
D	B
E	AC
F	AD
G	ED

Construct the network corresponding to this project.

9. Construct a project network in which some activity has an independent float that has a negative value.

Interpret this value.

REFERENCES

Eisner, H., 1962. A Generalized Network Approach to the Planning and Scheduling of a Research Project, *ORSA*, 10, pp. 115-125.

Elmaghraby, S., 1964. An Algebra for the Analysis of Generalized Activity Networks, *Mgmt. Sci.*, 10, no. 3, pp. 494-514.

Ford, L. R., and D. R. Fulkerson, 1962. *Flows in Networks*, Princeton Press, Princeton, pp. 151-161.

Kelley, Jr., J. E., 1969. Critical-Path Planning and Scheduling: Mathematical Basis, *ORSA*, 9, pp. 296-320.

Moder, J., and C. Phillips, 1970. *Project Management with CPM and PERT*, Van Nostrand Reinhold Company, New York, 2nd ed. (This is an excellent comprehensive introductory treatment of project networks.)

Pritsker, A. B., 1977. *Modeling and Analysis Using Q-GERT Networks*, Halsted Press, New York.

Pritsker, A. B., and W. W. Happ, 1966. GERT: Graphical Evaluation Review Technique; Part 1. Fundamentals, *J. of Ind. Eng.*, 17, pp. 267-274.

Pritsker, A. B., and G. Whitehouse, 1966. GERT: Graphical Evaluation Review Technique, Part II, Probabilistic and Industrial Engineering Applications, *J. of Ind. Eng.*, 17, pp. 293-301.

INDEX

Activity
  critical, 320, 330
  defined, 319-320
  dummy, 321
  time, 332
Algorithm
  assignment, 275
  Dantzig shortest path, 58-60, 79-81, 241, 258, 298
    generalized, 64, 75-78
  defined, 4
  Dijkstra shortest path, 43-45
  double-sweep, 67-72
  earliest arrival flow, 139-142
  earliest event time, 323-324
  flow augmenting, 91-93, 97
  Floyd shortest path, 52-54, 79-81, 241, 258, 298
    generalized, 64, 75-78
  Ford shortest path, 49-51, 96, 117
  greedy, 23
  K-th shortest path, 64-78
  latest event time, 327-328
  maximum branching 22-35
  maximum cardinality matching, 185-196
  maximum flow, 95-99, 109, 124
  maximum weight matching, 200-209
  minimum cost flow, 105-116, 132-133, 140, 246-247, 252, 273
  minimum weight covering, 214-221
  mixed postman, 251-253
  other shortest path, 78-81
  out-of-kilter, 116-126, 335

[Algorithm]
  simplex, 14
  spanning tree, 19-26
Arborescence
  defined, 26
  root of, 26, 276
  in salesman problem, 275
  shortest path, 45
  spanning, 27
  weight of, 27
Arc
  backward, 90
  capacity, 88, 117
  defined, 2
  forward, 90
  increasable, 88
  length, 41
  multiple, 48
  reduceable, 88
  undirected, 4
  weight, 27

Basic solution, 14
Bottleneck, 78
Branch and bound technique, 278-279
Branching
  algorithm, 29-35
  defined, 27
  spanning, 27
  weight of, 27

Center
  absolute, 297, 305-306
  defined, 290, 291, 298
  general, 297, 299-300
  general absolute, 297, 300-301
  multi-, 315-316

353

Chain
 alternating, 189
 augmenting, 189, 200-201
 defined, 5
 flow augmenting, 89-90
 length of, 5
 simple, 5
Chinese postman (see Postman
   problem)
Christofides, N., 275
Chvatal, V., 271
Circuit
 defined, 5
 Euler, 236, 238
 Hamiltonian, 261
   existence of, 265-267
 negative weight, 51
 salesman, 261
 shrinking of, 29-31
 simple, 5
Complementary slackness conditions
 defined, 12
 in flow with gains algorithm, 155, 161-162
 in maximum weight matching algorithm, 203
 in minimum cost activity times, 330
 in minimum cost flow algorithm, 110-113
 in minimum weight covering algorithm, 215
 in out-of-kilter algorithm, 120-121
Component
 connected, 6, 15, 19-20
 strongly, 265
Computational complexity
 of double-sweep algorithm, 72-73
 of generalized shortest path algorithms, 77-78
 of shortest path algorithms, 62-64
Constraint
 linear programming, 10, 11
Convolution, 75
Covering
 defined, 181
 matching generated by, 184

[Covering]
 maximum cardinality, 184
 maximum weight, 184
 minimum cardinality, 183
 minimum weight, 183
Cramer's Rule, 14
Critical Path Method, 318-331
Cut
 capacity of, 98
 defined, 7
 proofs using, 98, 126, 135
 simple, 7
Cycle
 absorbing, 156
 defined, 5
 even, 186
 generating, 155
 odd, 187
 shrinking of, 195, 205, 218
 simple, 5

Dantzig, G., 52
Decision box, 343
Degree, 15, 237
 inner, 15, 237
 outer, 15, 237
Dijkstra, E., 43
Distance
 point-arc, 294
 point-vertex, 292
 vertex-arc, 292
 vertex-vertex, 292
Dual linear programming problem, 10, 110-111, 124, 166, 206-207, 219, 337

Edge, 4, 19
Edmonds, J., 26, 102, 189
Euler, L., 1, 235-236
Euler tour, 236, 238
Event, 323
 algorithm for numbering, 323-324

Finite termination
 of double-sweep algorithm, 71-72
 of flow with gains algorithm, 169-172
 of maximum flow algorithm, 102-104

… INDEX

[Finite termination]
  of minimum cost flow algorithm, 113
  of out-of-kilter algorithm, 126-127
Float
  free, 329
  independent, 330
  total, 329
Flow
  augmentation, 89-91
  canonical, 169-172
  defined, 87
  dynamic, 128-129
  earliest arrival, 137-138, 149
  earliest departure, 146-150
  latest arrival, 146, 150
  latest departure, 146, 149
  lexicographic, 150
  maximum, 95
  minimum cost, 105
  units of, 88
  with gains, 151-153
  without gains, 159
Floyd, R., 52
Ford, L., 49, 102, 106
Fulkerson, D., 96, 102, 106, 117, 333
Forest, 7

Gain, gain factor, 80, 152
Gauss-Jordan elimination, 14
Generalized addition, 65
Generalized minimization, 65
Ghouila-Houri, A., 267
Graph
  bipartite, 185-187
  connected, 6
    strongly, 265
  defined, 2
  even, 237
  inverse, 147-148, 172-173
  matrix of, 8
  mixed, 249
  symmetric, 237
  time expanded replica, 129
  undirected, 4

Hakimi, S., 301
Hamilton, Sir W.R., 261

Independence, 333, 348

Input, 343
Interior point, 291

Jewell, W. S., 160
Johnson, E., 102

Karp, R., 102
Kilter number, 122
Konigsberg bridge problem, 1, 236, 259

Lexicographic preference, 150
Linear programming, 8-14
  flow with gains as, 153-155
  maximum cardinality matching as, 187-188
  maximum weight matching as, 202
  minimum cost activity times as, 324
  minimum cost flow as, 106-107
  minimum weight covering as, 214-215
  out-of-kilter problem as, 119-120
  primal, 11
Location theory, 289
Loop, 4

Matching
  covering generated by, 184
  defined, 181
  maximum cardinality, 182
  maximum weight, 182, 241
  minimum cardinality, 183
  minimum weight, 183
Matrix
  of a graph, 8
  lower triangular, 66
  shortest path length, 52, 58
  upper triangular, 66
Median
  absolute, 298, 308-309
  defined, 290, 306
  general, 298, 307
  general absolute, 298, 309-312
  multi-, 316
  weighted absolute, 298

Network
  defined, 4, 88, 97
  generalized project, 342

[Network]
  project, 320
Nonnegativity conditions, 10

Objective function, 10
Output, 343

Path
  critical, 331
  defined, 5
  flow augmenting, 89
  K-th shortest, 64-78
  length of, 41
  optimum, 66
  shortest, 41
  simple, 5
Point, 291
Postman problem, 235-258
Program evaluation review technique, 332-333

Salesman problem, 261-287
Sink, 88
  multi-, 104
Slack variable, 12, 13
Source, 88
  multi-, 104
Subgraph, 6
Successive improvement technique, 281-282

Traveling salesman problem
  (see Salesman problem)

Tree
  alternating, 191-192
  defined, 5
  Hungarian, 192, 195, 206
  spanning, 5, 19-26
Triangle inequality, 263
Truemper, K., 158

Variance, 333
Vertex
  adjacent, 5
  artificial, 30, 195-196, 205-207, 216-218
  defined, 2
  empty, 216
  explosion of, 118
  exposed, 189
  indicence, 4
  initial, 5
  inner, 191
  matched, 189
  outer, 191
  penultimate, 57-58
  saturated, 216
  strongly connected subset of, 265
  terminal, 5

Weight
  arborescence, 27
  arc, 19
  branching, 27
  covering, 183
  matching, 182

EXXON CORPORATION
MCS TECHNICAL LIBRARY